Let Thy Food Be Thy Medicine

LET THY FOOD BE THY MEDICINE

Plants and Modern Medicine

Kathleen Hefferon

OXFORD
UNIVERSITY PRESS

OXFORD
UNIVERSITY PRESS

Oxford University Press is a department of the University of Oxford. It furthers the University's objective of excellence in research, scholarship, and education by publishing worldwide.

Oxford   New York
Auckland   Cape Town   Dar es Salaam   Hong Kong   Karachi
Kuala Lumpur   Madrid   Melbourne   Mexico City   Nairobi
New Delhi   Shanghai   Taipei   Toronto

With offices in
Argentina   Austria   Brazil   Chile   Czech Republic   France   Greece
Guatemala   Hungary   Italy   Japan   Poland   Portugal   Singapore
South Korea   Switzerland   Thailand   Turkey   Ukraine   Vietnam

Oxford is a registered trade mark of Oxford University Press in the UK and certain other countries.

Published in the United States of America by
Oxford University Press
198 Madison Avenue, New York, NY 10016

Library of Congress Cataloging-in-Publication Data
Hefferon, Kathleen L.
Let thy food be thy medicine : plants and modern medicine / Kathleen Hefferon.
    p.   cm.
Includes bibliographical references and index.
ISBN 978-0-19-987397-5 (hardcover : alk. paper)
1. Materia medica, Vegetable. 2. Herbs—Therapeutic use. 3. Ethnobotany.
I. Title.
RS164.H335   2012
615.3'21—dc23   2011050053

9 8 7 6 5 4 3 2 1

Printed in the United States of America
on acid-free paper

CONTENTS

1. Plants and Human Health   3
    Our Early Relationship with Plants   5
    Plants as Medicine   12
2. Bioprospecting for Medicines from Plants   14
    Ethnobotany and Medicine   16
    Modern Drug Discovery and Indigenous Cultures   20
    Plants and Modern Drug Discovery   23
    Examples of Medicines Derived from Plants   27
    Impact of Biopiracy, Preservation of Biodiversity   37
    Intellectual Property Rights for Indigenous People   38
    Conclusions   41
3. The Lure of Herbal Medicine   43
    History   45
    The Bioactive Compounds in Herbal Medicinal Plants   50
    Traditional Indian Medicine   56
    Traditional Chinese Medicine (TCM)   62
    What Nonscientists Should Know about Herbal Medicines   71
    What Scientists Need to Know about Herbal Medicines   75
    Could Herbal Medicine and Western Medicine Complement Each
    Other?   77
    Conclusions   79
4. Farming Medicines from Plants   82
    Why Farm for Pharmaceuticals in Plants?   83
    New Production Systems   86
    How Does It Work?   87
    Technologies Used to Design Plants Expressing
    Biopharmaceuticals   89
    Clinical Trials of Therapeutic Proteins Produced in Plants   97
    Allergies, Oral Tolerance, and Dose Response Relationships to
    Plant-made Vaccines   101

The Scale-Up and Commercialization Opportunities for Plant-derived Therapeutic Proteins *102*
Plant Production Platforms for Molecular Farming *104*
Molecular Farming Requires Its Own Unique Set of Regulatory Guidelines *104*
Conclusions *105*
5. Superfood: Functional and Biofortified Foods *107*
The Science behind Functional Foods *108*
Functional Foods and Human Genetics *111*
The Mediterranean Diet: The Ideal Diet? *112*
Do Dietary Supplements Work? *121*
Functional Foods as Superfoods? *122*
Biofortified Foods and Hidden Hunger *123*
Can Biofortified Foods Make a Difference? *132*
Conclusions *133*
6. Food Security, Climate Change, and the Future of Farming *135*
Norman Borlaug and the Green Revolution *137*
The Green Revolution Missed Africa *142*
Agricultural Makeover: Sustainable Intensification *145*
Up–and-Coming Technologies: Crop Improvement *146*
Improved Farming Techniques *151*
Environmental Uses for Plants *155*
Agricultural Sustainability and Organic Food *158*
Conclusions *162*
7. The Future *164*
Foods as Drugs *165*
What Will the Food Industry Look Like? *168*
An Urban Answer to Increased Food Production: Vertical Farms *169*
Agricultural Sustainability and Protection of Biodiversity: A Daunting Challenge *171*
Biodiversity and the Search for New Medicines *172*
The Importance of Public Perception in Shaping Our Future *174*
Regulating Agricultural Innovations Can Be a Double-Edged Sword *175*
Conclusions *177*

Notes *179*
References *181*
Index *191*

Let Thy Food Be Thy Medicine

CHAPTER 1

Plants and Human Health

On a recent trip to China, I visited one of my PhD colleagues from graduate school, a Canadian with Chinese roots who saw a business opportunity to start a biotech company in his country of birth. Ever the entrepreneur, he told me of his plans to examine and evaluate a Chinese plant that he thought showed great potential as a new drug. The plant in question has thrived in the fields of his ancestral home as far back as any of his family can remember, and it has been used for generations as a general folk remedy and disinfectant. My colleague felt positive that we were looking at a plant that could have enormous value as a novel form of antibiotic, and perhaps could even offer a solution to the problem of multiple drug-resistant bacteria experienced by hospitals all over the world. I was a little reluctant at first about the concept of using natural compounds as medicines; it seemed in some way to be counterintuitive to my scientific background in biotechnology. My colleague, known among his friends to be a shrewd and rather good plant biochemist, seemed nonplused about my apparent misgivings. A search through available scientific publication and herbal medicine databases yielded no evidence that singled out this particular plant as possessing any known medicinal properties that had been recorded previously. As far as we could tell, we were entering uncharted territory. I gradually gave in to my curiosity and decided to follow my colleague on this "bioprospecting" expedition.

The first step was to see whether the results that my colleague anticipated could be reproduced in a laboratory setting. The initial experiment was simple enough to perform. We ground the plant up into a powder, dissolved this powder in a variety of different chemical solvents, and tested these extracts on different strains of bacteria. We found to our delight (and my amazement) that some of the extracts stopped bacterial growth completely! The next rational step seemed straightforward enough in principle: could we identify what active compound(s) in the plant is(are) responsible for inhibiting the bacteria? From a practical standpoint, this is not necessarily as easy as one would imagine. Biological molecules can be intrinsically complex, and although we could think of some logical directions in which to proceed, it seemed unlikely that we were going to come up with a tangible result for quite some time.

With great enthusiasm, nonetheless, we began to discuss the implications of our results more thoroughly. Had we discovered a new medicine? If so, would it be wise to ensure that our discovery was protected by a patent? Had we tapped into a gold mine? That point brought up more concerns. Did we have a right even to be thinking in terms of value and intellectual property? We may have been the first to test this plant using a standard scientific approach, yet the plant in question had been known by the locals of the region to have medicinal properties for as long as anyone could remember. Surely these people should have some right to this discovery.

Thinking along the lines of intellectual property, it was entirely possible that we had merely identified an already known natural compound, only from a different plant source. Had we actually "rediscovered" a chemical compound that wasn't novel at all, but perhaps was present in other plant species and had already been developed and marketed as a drug or a herbal medicine? The answer wasn't clear.

The more we dwelled on our discovery, the more sober our thoughts became. What if the plant component, which worked so well on a Petri dish (under in vitro conditions), worked much less convincingly on an actual person who suffers from bacterial infection (under in vivo conditions). It's true that there are many unofficial claims of the plant's ability to cure people of infection, but no actual documented proof of this existed. The only available evidence was from the affirmation of the local community. What if the plant had adverse effects on people or interacted with other medicines that one might be taking concurrently? How much would one need to take in a single dose and how would it be administered? Would it be provided as a tea or as a pill, for example? There must be some regulatory format that we should follow to ensure that our plant compound is unique, works in the way that we predict, and is safe to use. It seemed to me that surely all of these questions needed to be addressed before it was time for my colleague and me to celebrate.

Our thoughts turned to existing herbal medicinal products that are available on the market today. What about these sorts of products? Have the active components been identified? Do they actually work and are they safe? Do we really know that much about them at all? Furthermore, are any of the plants or other natural products used derived from endangered species? Does the process of collecting specimens impact the biodiversity of the region? Are there laws in place to protect against environmental damage from people collecting these plants? My colleague assured me that the plant we used in our initial experiment was common enough in the area, and the sample we had used for our experiments was from his family's garden. Our circumstances seemed harmless enough, but surely that is not always the case for the sampling of other natural bioactive compounds.

Other, more philosophical questions came to mind as we continued to discuss the project. What is the relationship between the traditional use of medicinal plants and modern medicine today? What role will plants have with respect to human health in the future? It was an attempt to answer these sorts of questions that led to the premise for this book. The search for the answers was definitely a journey to strange and unusual places. Beginning with our Paleolithic ancestors, this book moves on to cover our complex and fascinating involvement with plants, as both foods and as sources for medicines. The pathway leading from natural plant products to modern drug discovery is discussed. The delicate balance between maintaining biodiversity, the rights of indigenous tribes, and the identification and exploitation of novel bioactive compounds is examined. This book touches on the general spirituality associated with plants used as medicine, ranging from ancient Confucian and Ayruvedic philosophies to actual connections with witchcraft and on to widespread beliefs continued today in African cultures and others. The ways that traditional and conventional medicines merge and complement one another are discussed. The book continues with a glimpse into the changing yet equally important role of plants as novel innovative medicines and as "functional foods" that provide preventive measures against chronic diseases. This book underscores the importance of devising new ways to think about plants and agriculture in general, not only to address a soon-to-be burgeoning world population but also to help us to navigate our way through the impact of climate change. The book concludes with a projection of the role of plants in human health for next 100 years and presents possible means by which plants can play a predominant role in shaping our future.

OUR EARLY RELATIONSHIP WITH PLANTS

How were plants involved in our development into the beings we are today, in terms of the shape of our digestive tracts, our select dietary niches, and even the evolution of our social communities? It has become perfectly clear that a plant-based diet can play a role in helping us to avoid many of today's chronic health problems such as diabetes, high blood pressure, stroke, and some cancers. Did early hominids encounter the same chronic diseases? Do our diets today in some way fail to meet our true nutritional needs? Can crops be designed and bred to improve our health?

The overwhelming majority of primates consume plants as the substantial part of their diet; animal matter constitutes only a small proportion of their food source. Early primates, including man, most likely ate a wide number and diversity of plants that resided in their tropical rainforest homes; these happen to be dicotyledonous plants and would include the ancestors of many fruits, roots, and leafy vegetables that we consume today. Over time, early hominids left the forest and entered savannah regions, where the principal plant species to be found are monocotyledonous, such as cereals and grains. Were our digestive tracts adaptable enough to make the switch to these less traditional plant foods?

In addition to plants, many primate species supplement their diet with some animal matter, including insects and small mammals or birds. The digestive tract of some primates largely reflects these omnivorous food choices; humans, for example, tend toward a larger volume of small intestine and smaller colon than do gorillas, strict consumers of leafy plants, who exhibit the inverse ratio of gut proportions.

Clues with regard to the dietary niches of early humans are scant in evidence and high in conjecture. The reconstruction of prehistory can be ambiguous; and it is difficult to make any broad conclusions. Data have been collected from a variety of sources, including the dietary patterns of closely related primates, early hominid tooth micro-wear patterns, analysis of bone growth, density, and other skeletal pathologies. Together, much of this does not provide an adequate representation of what actually took place; rather, these data symbolize a small sample size of what has been archaeologically preserved.

Upon entry into the savannah setting, early man likely had to cover large ranges of land in search of food. The focus of diet changed to large game, which was complemented by vegetable-like foods. In their continuous search for food, early hominids became bipedal in terms of locomotion. What motivated the switch from largely a plant-based diet to one composed of animal meat and fat? The hunter-gatherers of temperate to Arctic regions of the world seem to have followed this transitional pathway, perhaps in the need for more calorie-rich foods, or perhaps the increased fat intake derived from eating game helped them to survive and flourish in colder climates. As humans became more proficient hunters of other animals, they changed their behavioral patterns as they developed new strategies by which to capture their prey.

The Paleolithic Age then led to the Neolithic Age, when the hunter-gatherer became an agriculturalist. It is still unclear what led to the transition.

Some experts attest that a rapid increase in population led to a strain on the customary food supply; others have suggested that changes in climate may have altered the availability of animals to hunt as meat. In the Middle East, these changes in climatic conditions led to the favored growth of grassy plants such as wild wheat, rye, and barley. Faced with a dwindling supply of animals to hunt, early man naturally turned to grain cultivation. Doing so meant a significant change with respect to lifestyle. Gathering wild grain when it ripens and loses its seed to the wind requires large groups of people to be available at the right place and at the right time to catch the harvest. This in turn would require the development of settlements to be at the ready for the time of harvest. Using a scenario such as this, the hunter-gather may have morphed into the first farmer.

At first, this may have involved at most the smallest of efforts, for example, by cutting back the vegetation so that a patch of a particular species of plant that was good to eat could receive more sunlight and the space to thrive and grow. Other primitive methods involved the selective burning of old growth vegetation to encourage the rapid growth of tender edible leaves and shoots, either for human consumption or to sustain grazing for large game animals in proximity to the hunter-gatherers. Plant tubers and roots that were found to be tastiest and most nutritious were most likely collected; a few were transplanted and left to propagate in designated areas to ensure a future food source. Such forms of land management by early "proto-farmers" were practiced in one form or another more than 10,000 years ago by many of our Neolithic ancestors throughout the Americas, the Middle East, and Asia.

It was only a matter of time before what can only be called the first true plant breeder began to emerge, as early man learned crop cultivation as a means of ensuring reliable future harvests. One possible scenario that led to this revolution in our history is the collection and preservation of select seeds at the end of each growing season. The act of selecting seeds to be spread at the beginning of each new season most likely took place at a subconscious level at first, in response to the quality of each season's crop. Early farmers learned that scattering some of the seeds they had collected from their best harvests ensured them of more of the same for the next year's growing season. Perhaps the particular seeds that the first farmers chose to be saved for future sowing produced food crops that had larger yields, were tastier, or ripened more readily. This selection of seeds from the most suitable crops was man's first attempt at deliberately breeding for the traits he desired, and thus altering the course of gene flow for a given population. It is safe to say that the predictability of this arrangement soon led to the establishment of a co-dependence of man and plant. The end result was the domestication of crop plants and the emergence of

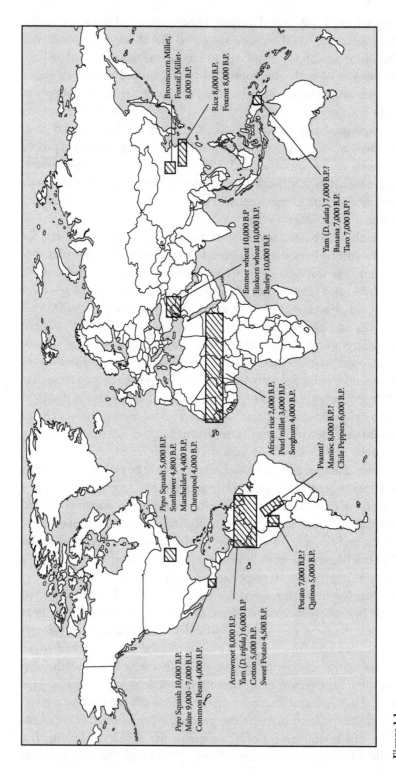

Figure 1.1.
Currently recognized independent centers of plant domestication. From *Proceedings of the National Academy of Sciences, U.S.A.* 103, no. 33 (August 15, 2006): 12223–28.

early inhabitants into what we recognize today as civilized society, one in which a variety of occupations in addition to agriculture were required so that the settlement as a whole could function smoothly.

Many of the crops the world uses today as staples were first domesticated by approximately 6000 BC (Figure 1.1). Man's active participation in breeding for select traits was not sufficient to make early crops such as barley or wheat a staple in the diet, however. It was clear that raw grain itself was indigestible; some processing mechanism had to be set in place to thresh and grind the grain, store it for future use, or cook it so that it was suitable for eating. Irrigation systems were devised to ensure a consistent water supply and thus reliable harvest in times of drought. Trading routes were established between different settlements, so that a variety of foods, and later other material goods, could flow to distant lands. It is around this time that distinct cultures began to materialize (Figure 1.2).

While large fields of grain were becoming the first domestic staple crops in the Middle East, early farmers in many other parts of the world were plotting small gardens containing a wide assortment of fruit trees, vegetables, herbs, spices, and a variety of medicinal plants. In some of these primitive societies, crop plants were attributed a supernatural reverence, perhaps even viewed in some instances as "gifts of the gods," as is the case of corn in Mayan cultures. Plants that were used for medicinal purposes were often associated with an aura of mysticism, and treatment of the sick involved distinct healing rituals performed by an appointed individual who was believed to be a link to the spirit world. The relationship between plants and health as it relates to dietary habits and therapeutic choices of healing remains richly steeped in custom, taboo, tradition, and spirituality even to this day.

An eclectic mix of herbalism with spirituality is showcased in the health practices of many cultures that exist in the present, such as traditional African medicine. Distinct from the reductionist, analytical approach of Western medicine, traditional African medicine takes into account the role that ancestral spirits have on an individual and the surrounding environment as direct links to people's health and well-being. With the number of Western medical doctors too small to address the current population, traditional healers and remedies made from indigenous plants remain critical to the health of millions of Africans to this day (Figure 1.3). Furthermore, modern pharmaceuticals are often hard to come by or are too costly. Indeed, over 80% of the world's people rely on herbal remedies rather than modern drugs, and traditional herbal practitioners are their only means of medical expertise.

The evolutionary path from Paleolithic to modern man sounds straightforward enough, but nutritionally speaking, nothing is that simple. While

Figure 1.2.
Agricultural workers in ancient Egypt.

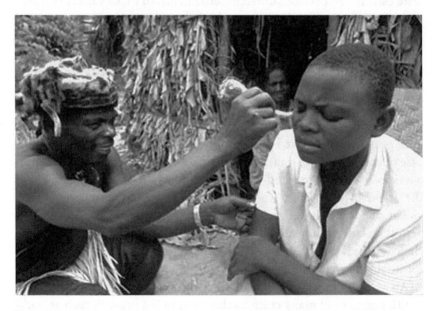

Figure 1.3.
A traditional healer in Uganda treating a patient's dizziness. From Africa Renewal, United Nations.

one would assume that cultivating crops, consuming more plant and less animal matter, settling into towns and cities, and establishing sophisticated trading routes must directly indicate improved health of man in general, a proportion of data suggests otherwise. A comparison of skeletal pathologies of hunter-gatherers and farmers suggests that farmers were in general more prone to infection, had a shorter life expectancy, and experienced more chronic malnutrition than their hunter-gatherer counterparts. It seems, according to this argument, that substituting a meat-based diet for one that is grain-based had a detrimental impact on our overall health.

Often these data have been used to support the popular "Paleo Diet," a high meat, low carbohydrate diet that is embraced by many today as what our true eating lifestyle should be. The reasoning is that since we have the same genetic composition as our hunter-gatherer ancestors, our style of life and dietary choices should mimic theirs. Our current diet contains plant-based foods high in carbs, such as breads, potatoes, rice, and pasta; these were introduced only recently as food sources and thus are associated with a number of modern diseases we face today, such as diabetes, heart disease, and some cancers, many of which can be directly linked to diet. Did these diet-related diseases increase with the evolution of agriculture?

While a high-carb diet and sedentary lifestyle have been blamed for many ills of modern society, several points contradict the assumption of better health for our earlier ancestors. First, although the sequence of our genes matches those of our ancestors, our epigenetic profile, or the way these genes are expressed, may differ substantially. In addition, our civilization has brought with it a life expectancy of over 70 years, an age unheard of in prehistoric times. Overconsumption of meat and fat makes post-reproductive individuals more prone to succumb to heart disease at later stages of life; however, the life span of Paleolithic man was not long enough to experience these conditions. As Reay Tannahill puts it in *Food in History,*

> On the principle that twentieth-century diseases are largely a product of the twentieth-century diet, they recommend a return to the foods of our ancestors (dates unspecified), who did not die—as so many people do today—of cardiac thrombosis, strokes, or cancer.
>
> This is perfectly true. Our ancestors died, instead, of malnutrition, diabetes, yaws, rickets, parasites, leprosy, plague, skin infections, gynecological disorders, tuberculosis, and bladder stones, and they usually died in their 30s. Most modern diseases do not develop until the victim is in their 40s or 50s. If our ancestors lived 10 years longer, coronary thrombosis, strokes and cancer might have been their fate too.[1]

So the dogma for a diet high in meat is not without fault. Alternatively, one can equally argue that a hazard of strict vegetarianism is the difficulty in securing sufficient nutrients for optimal health. Arguments regarding the benefits and pitfalls of animal- versus plant-based diets are engaging, and the taboos associated with food throughout various cultures are fascinating in their own right. This book, however, is about plants and their current role in human health, and it is with this brief history of man's beginnings that this book commences.

Plants, unlike most animals, are immobilized to one place; they cannot flee or fight off their predators. Instead, they have devised different means to avoid being eaten; they can be thorny, bitter to taste, and poisonous.

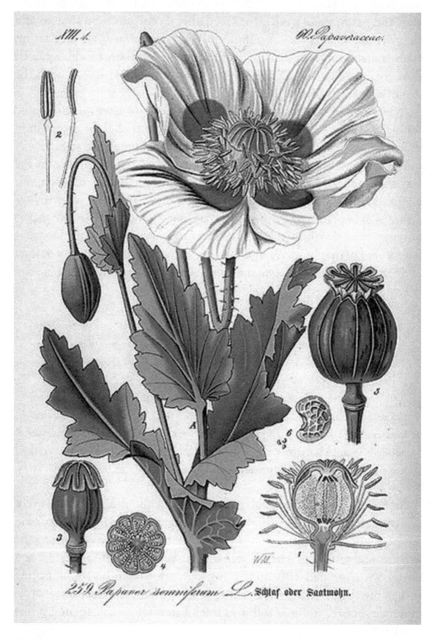

Figure 1.4.
(a) Opium poppy (*Papaver somniferum*).

Figure 1.4.
(b) A Chinese opium house, photograph circa 1900.

They can contain anti-nutrients, compounds such as phytates that prevent valuable minerals from being absorbed in the digestive tract, making them less attractive as a food source. Any and all of these features help to deter them from being eaten, and foragers are warded away from consuming specific plant species that exhibit these attributes. Oddly enough, it's among plant chemicals, or phytochemicals, that so many of our medicines are found. Among the plant-derived drugs used as medicinal compounds throughout history and even today are morphine and digitalis. For example, the pain reliever morphine is produced from the opium poppy, as are heroin and codeine (Figure 1.4).

Many plants possess bioactive compounds that can help ward off chronic illnesses, such as cardiovascular disease and even some cancers. Lycopene, a compound found in tomatoes, and anthocyanin, found in blueberries, are examples of anti-oxidants—plant compounds that are abundant in a wide assortment of fruit and vegetables. The use of plants as functional foods to prevent chronic diseases is outlined in detail in Chapter 5. Different plant compounds can even be expressed in different anatomical parts of the plant, such as the roots, leaves, and flowers. Alkaloids, for example, can be extremely poisonous to humans but a number, like morphine, atropine, and cocaine, are widely used in medicine. As the next chapter will reveal, bioactive compounds derived from plants and other forms of nature lie at the root of modern drug discovery.

CHAPTER 2

Bioprospecting for Medicines from Plants

The field of pharmacology traces its origins from our herb-gathering ancestors. Plants have been intimately associated with medicine from the beginning of humanity. Early civilizations frequently attributed a mystical quality to the world around them. Injuries and diseases were often considered the result of the actions of evil spirits who were able to directly influence people's lives. It is not difficult to imagine the sense of absolute wonder that early humans must have experienced upon observing the healing characteristics of certain plant species. For highly impressionable people of primitive cultures, the identification of plants that possess some form of healing property, whether real or invented, must have appeared as magic. It comes as no surprise then, that much of early medicine maintains its foundations in the supernatural and was practiced by a tribal member designated as the contact point with this other world. In primitive societies, the healing process took place in a highly orchestrated ceremony, brimming with rituals and shrouded in mystery, in part to reassure the patient that the treatment would in fact be effective. For many traditional shamans or healers across the world and throughout history, the identification and use of medicinal herbs were believed to be directed by ancestral spirits. This spiritual guidance reaffirmed a strong and complex relationship between the species of plants used for medicinal purposes, the ritualistic manner in which they were administered, and the belief in the powers of the departed over the living. From these humble beginnings developed the intriguing history of plants and their association with the field of pharmaceutical drugs as we know it today.

The term "ethnobotany" comes from "ethnology," the study of culture and "botany" the study of plants, and refers to the interactions that exist between plants and select cultures of indigenous people. These populations hold the key to the identification of plants that may contain bioactive compounds possessing some pharmacological value. Several approaches have been used to search for and identify novel bioactive chemical entities from natural sources. One consists of the random collection and screening of many different species of plant material, and this was the strategy used to identify Taxol, an effective anti-tumor agent derived from the Pacific yew tree *Taxus brevifolia*. Another approach involves the compilation of ethnopharmacological knowledge from indigenous peoples of a particular region, and then the use of these as leads in the search for new medicines. An example of this strategy is the identification of the antimalarial agent artemisinin from the sweet wormtree *Artemisia annua*, which has been used in Chinese medicine since 200 BC.

Chemical compounds originating from the forests of the world today serve as the framework from which scientists develop novel drug compounds in the laboratory. The wealth of the natural environment presents a fundamental research base for the field of drug discovery and has brought renewed interest in the preservation of biodiversity, since bioactive compounds found in the wild may not be reproducible in the laboratory. A significant proportion of medicines available today is based on plants and microbes, or synthetic chemicals derived from plants and microbes. The development of biotechnology and high throughput screening techniques enables large numbers of plant samples to be screened in a short period of time. These technologies have enhanced the search for novel bioactive compounds in various plant species, by combining the fields of plant natural product chemistry and ethnobotany.

Only a small fraction of Earth's total biodiversity, including plant, animal, and microbial species, has yet been tested for novel biological activities. Organisms that live in a species-rich environment have evolved a multitude of defensive and competitive mechanisms to survive, some of which may be exploited for the benefit of mankind. Plants within these habitats frequently possess a great diversity of chemical compounds which they use to protect themselves from viral and fungal pathogens, as well as animal predators, and to compete with other plants for the same resources, light and nutrients. In certain cases, indigenous populations can offer advice as to which plant or plant part may be most effective in treating an illness or disease condition. More often than not, the regions of the world that harbor the greatest degree of biodiversity lie in tropical climates. Unfortunately, these areas also tend to be poorly environmentally protected and sometimes even politically unstable, making them less than ideal candidates for sustaining ecosystems that may house the world's next wonder drug. With poor regulatory infrastructure and implementation, countries that happen to be located within these regions are threatened with respect to biodiversity conservation and are easily subject to unsavory practices such as biopiracy.

This chapter describes the search for novel medicines from plants using ethnobotanical knowledge combined with the practices of modern drug discovery. It traces the step-by-step progression that a plant with potential pharmaceutical properties must take through research and development to the final stages of clinical trials and regulatory release. Examples of plants that have been used in the past for well-known and highly successful medicines are provided, as are plants currently under investigation as potential new leads for drug discovery. The chapter ends with a discussion of the current status of intellectual property rights for indigenous peoples and the urgent need for preserving biodiversity in these vital parts of the world.

ETHNOBOTANY AND MEDICINE

Many approaches have been used to identify plants that may possess medicinal properties. One practice is to use structured interviews to examine the oral tradition of indigenous populations where the plant grows. The transfer of information through the oral tradition is not unusual for many indigenous peoples; for example, the selection and manner of use of plants for medicinal purposes may be passed down from one healer or shaman to a family member or apprentice of the next generation. It is unfortunate that surprisingly little attention has been paid to maintaining and recording the wealth of information available through oral transmission in indigenous populations. Even though oral-based knowledge systems regarding indigenous medicinal plants are rapidly disappearing, written recordings of this information have only rarely been the focus of research projects.

How are decisions made regarding the use of plants for medicinal use by indigenous populations? In some cultures, it can be partially a matter of remembering a plant's smell and taste. Some cultures use these chemosensory perceptions to determine the therapeutic properties of a particular plant species as well as how best to prepare and administer them. Chemical compounds within the plants, such as secondary metabolites, often have a distinctive smell or taste that provides clues to their potential therapeutic value. In addition to this, the therapeutic uses derived from traditional knowledge are often intermingled with ceremonial and symbolic ones.

Today, much of the ethnobotanical knowledge relating to medicine that does get recorded is derived from various geographical regions that are considered to be "hot-spots of biodiversity" and include localities ranging from South America to Southeast Asia. For example, a series of villages have been used as study sites for acquiring knowledge of medicinal plants from indigenous people in the eastern part of the Shimoga district of Karnataka State in India (Figure 2.1). Information about the species of plant

a   b

Figure 2.1.
(a) Map of Karnataka. From mapsof.net; (b) Shola grasslands and forests in the Kudremukh National Park, Western Ghats, Karnataka. Photo by Karunakar Rayker.

used to treat a particular illness as well as the preparation of plant parts, dosage, and mode of application was collected and documented by conversing with local herbal practitioners as well as various individuals who received herbal treatments. A database of plant species was then constructed and further categorized according to whether a particular plant species was frequently used for treating a single disease category or whether there was less agreement among the informants about its use to treat a particular disease.

Other valuable sources are ancient and medieval texts as well as diaries or travel accounts written by explorers who had a combined interest in botany and medicine. These databases can be searched for plants with potential medicinal properties. For example, approximately 50 different plant species have been continuously used for medicinal purposes in the Mediterranean region of Campania over several hundred years. Today, there are many available databases in which traditional knowledge on the use of plants from different geographical areas has been collected, stored, and maintained (Figure 2.2). One example is the International Ethnobotany Database, an online, multilingual database that can be accessed by the public (http://www.quantumimagery.com/software-ebDB.html). Another popular database is the Medusa Network (http://medusa.maich.gr/), which manages ethnobotanical information from informants, recent and ancient documents, actual specimens, and in some instances archaeological remains.

Ethnobotanical information can often be found in records kept by explorers on expeditions, as they would occasionally come into contact with isolated indigenous tribes. For example, the Khoi-Khoi (pastoralists for the

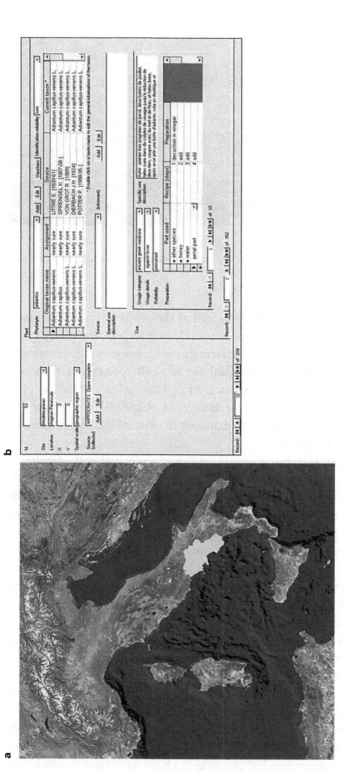

Figure 2.2.
(a) Location of Campania region in Italy; (b) Main data management mask, showing a record of ancient ethnobotanical data. A relational database has been built on the basis of information gathered from different historical sources, including diaries, travel accounts, and treatises on medicinal plants, written by explorers, botanists, and physicians who traveled in Campania during the last three centuries. Moreover, ethnobotanical uses described in historical herbal collections and in ancient and medieval texts from the Mediterranean Region have been included in the database. From *Journal of Ethnobiology and Ethnomedicine* 5, no.7 (2009), doi:10.1186/1746-4269-5-7.

last 2,000 years) and San peoples (hunter-gatherers in the region over 20,000 years ago) are believed to be among the earliest inhabitants of South Africa. Both groups of peoples have a robust and extensive ethnomedical background which is largely based on indigenous plant species. This knowledge is passed down from generation to generation through the oral tradition, and the culture of both tribes is under increasing threat of extinction. Written records of Khoi-San traditional medical practice are preserved in sources such as the journal entries regarding exploratory expeditions and archival records maintained from activities of the former Dutch East India Company (VOC) at the Cape of South Africa (Figure 2.3). Specimens of a number of these African medicinal plants were removed from their native environments and grown in herbariums in other parts of the globe, where some are still propagated. The archives of the Dutch East India Company are now considered an important source of ethnobotanical information. Select pieces of the archives have been recognized as having potential applications in both drug discovery and intellectual property protection. Besides the archives of the Dutch East India Company, South Africa also has well-kept ethnomedical records dating from early 1800 onward. These records are compiled in a single volume and represent an accumulation of both field studies and extensive detailed literature searches of plant species used as traditional medicines by all ethnic groups living in South African countries.

Figure 2.3.
Arrival of Jan van Riebeeck (of the VOC) in Cape Town painted by Charles Davidson Bell.

How has the modern pharmaceutical industry utilized the knowledge of indigenous cultures, largely inhabiting ecologically fragile areas and often in poor countries, to incorporate plant-derived compounds into their drug discovery programs? One popular approach is for the drug company to form some sort of strategic alliance with the government of the country harboring the indigenous population.

The most widely publicized bioprospecting project agreement between a commercial entity and a conservation group took place in 1991, between Merck and Co., one of the world's leading pharmaceutical companies, and the National Institute of Biodiversity (INBio) in Costa Rica, a nonprofit conservation organization. The purpose of the Merck-INBio collaboration was to search for naturally occurring therapeutic agents produced in plants, animals, and microbes within the country. INBio's interest in the agreement was to establish avenues that would preserve Costa Rica's diversity and prevent deforestation. The collaboration with INBio had several advantages. The Costa Rican government's reputation as a stable democracy added a sense of security for the collaboration. In addition, Costa Rica's diverse geography, encompassing both coastal and mountainous regions, represents a vast assortment of tropical ecosystems. Scientists estimate that Costa Rica has more biological resources per hectare than any other country on the planet (Figure 2.4).

Figure 2.4.
Biodiversity in Costa Rica. Photo by Sandy Wiseman (sandy.wiseman@utoronto.ca).

Merck consented to provide $1 million as well as technical support to INBio for the right to screen and analyze an agreed-upon number of indigenous plant and animal samples and promised royalties for any commercially viable product. A portion of the capital was to be invested in biodiversity conservation through Costa Rica's Ministry of Natural Resources, and the remainder used to preserve the environment at the discretion of INBio's board of directors. In addition, Merck agreed to establish research facilities and train Costa Rican scientists, thus contributing to that country's economy.

Similarly, the National Cancer Institute (NCI) has also taken steps to ensure that the plant materials that it gathers in its search for anticancer drugs and drugs to combat the human immunodeficiency virus (HIV) are handled in an equitable way. In this case, a Letter of Collection is prepared for researchers and host nations which recognizes the need to protect the interests of the countries from which the plant material was gathered.

A second tactic used to identify medicinal compounds in plants is that employed by Shaman Pharmaceuticals. Instead of mass screening tens of thousands of plant species, companies like Shaman speeded up the search by collecting information from healers and herbalists of indigenous cultures who used plants in common to treat specific ailments. Shaman then compiled a database to narrow down and focus on a few potential plant candidates in tropical regions of Asia, Africa, and Latin America. Furthermore, Shaman promoted the conservation of tropical forests and made a commitment of reciprocity to indigenous cultures. The long-term reciprocity involved returning a portion of the profits to the indigenous communities once a commercial product was realized. As deriving a marketable medicine from plants may take 10 years or more, Shaman founded the Healing Forest Conservancy, a nonprofit organization with an independent board of directors and advisors, as a mechanism by which to compensate indigenous people for their knowledge and contribution to drug discovery. Since the journey of developing a plant product into a new medicine is a lengthy one, this immediate compensation was intended to directly pay back indigenous cultures before a marketable product was developed. The Healing Forest Conservancy was dedicated to preserving cultural and biological diversity and to sustaining the development and management of the natural and biocultural resources that comprise a portion of the heritage of native populations.

Unfortunately, approximately 10 years and $90 million later, the company was not able to meet the requirements of the US Food and Drug Administration (FDA) to successfully complete Phase III clinical trials for its lead bioprospecting drug candidate, Crofelemer. This drug has appeal in the developed world as a treatment for diarrhea related to acquired immunodeficiency syndrome (AIDS) and HIV, as well as diarrhea resulting from irritable bowel syndrome, and in the developing world for pediatric diarrhea and acute infectious diarrhea. Crofelemer is derived from the *croton lechleri* tree, which oozes a red latex sap known as "dragon's blood" (Figure 2.5). Crofelemer acts by normalizing water flow in the gut. It works in the intestine to treat diarrhea and prevent dehydration. Eventually, Shaman Pharmaceuticals filed for bankruptcy. However, there is a silver lining to this story, as a new company known as Napo Pharmaceuticals, named after an Amazonian River and headed by Shaman's former CEO and several other core Shaman employees, was founded in 2001. Napo purchased the rights to Crofelemer through bankruptcy court. Napo holds supply rights for sustainable harvested raw material and intellectual property rights for the compound. In the fall of 2010, Napo completed its analysis of a Phase III study and expects to commercialize Crofelemer in 2012.

A third approach that has been used by bioprospecting companies is demonstrated by the San Francisco-based company Diversa. This company's business model hinges on a biotechnological approach. DNA isolated from various plant species of a given region is collected and cloned into easily manipulated expression systems. Massive numbers of clones can then be screened for potentially interesting products by sequence analysis. One attractive feature of this methodology is that the actual tissue quantity of each biological specimen required is minute. Since one is dealing here with a plant's genome, it is not necessary to collect a number of different

Figure 2.5.
Croton lechleri, or Dragon's Blood tree. Photo by Ricarda Riina.

plant parts for each species. One small sample will be sufficient to completely sequence the organism's genome and then to search for genes that may have potential medical use.

Diversa entered bioprospecting agreements with numerous institutions, including INBio of Costa Rica, Bogor Agricultural University in Indonesia, the Institute of Biotechnology at National Autonomous University in Mexico, and Yellowstone National Park in the United States. Although approximately 700 new enzymes were identified using this approach, only a handful of patents were issued, demonstrating the high risk involved in bioprospecting research.

PLANTS AND MODERN DRUG DISCOVERY

A great many traditional drugs used in Western medicine are derived from plants, so it is no surprise that plants remain an important source of starting material for drug discovery. Researchers have discovered a number of drugs either by identifying the bioactive compounds from traditional herbal remedies or by mass screening large numbers of plant species. A third approach that is recently being pursued involves first understanding how diseases are handled by the body at the molecular level, and then utilizing this knowledge to identify suitable drug candidates that can either block or activate a particular target molecule. This third approach is the fulcrum of modern drug discovery (Figure 2.6).

Drug discovery in essence first involves the identification of lead candidates to combat a specific disease, followed by their extraction from plant material or synthesis in the lab. Thus, "drug development" includes a complete characterization of the phytochemical to determine its mechanism of action, toxicity, and therapeutic efficacy. Once a compound has shown its value in these tests, it will be taken through a series of clinical trials to determine how toxic and efficacious it is in a human population. Successful completion of the various phases of clinical trials is critical for obtaining the necessary governmental approvals for release of the plant-derived drug to the general public. In the United States, for example, drugs must be approved by the Food and Drug Administration. Drug discovery is a lengthy and expensive process, with research and development costing between $1–2 billion and taking approximately 10–15 years from lab bench through clinical trials to a final product.

To search for and identify a new drug that will act against a specific target for a particular disease, researchers often use a highly automated technique known as high throughput screening. In this case, "libraries" consisting of vast numbers of chemicals are tested and thus "screened" for their ability to

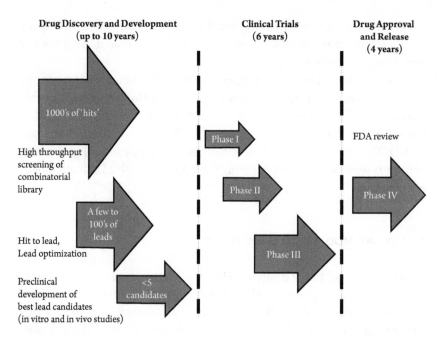

Figure 2.6.
Steps involved in drug discovery and development.

alter the target in a form of assay, or test, for a specific biochemical activity, such as inhibition of the function of a particular target molecule. Often, a natural compound derived from a plant that is identified as having some sort of drug-like activity against a particular target molecule is used as a "skeleton" for the compilation of a combinatorial library, with the hope that a slightly modified or "optimized" version of the original compound would improve the drug-like properties of the molecule. One reason for the use of natural products in this fashion is that they inherently possess a broad and robust chemical structural diversity. Made to order combinatorial libraries based on a particular skeleton molecule can be purchased commercially from chemical companies and are available in sizes as great as several millions of compounds. The compounds obtained from this high throughput screening process that show the most promising activity toward the target molecule are called "hits" and, upon repeated success, can be promoted to "lead compounds."

Once the ideal version of the natural compound is found, high throughput screening may be used to show how selective the compounds are for the chosen target, and to demonstrate that they do not interfere with other related targets or cause adverse side effects. At certain points in this process, new chemical libraries that act as extensions of the original are constructed and screened. Slight modifications may be made to the lead compound, such as the change of an end group, to facilitate recovery of the compound during synthesis or extraction. A new selection of compounds corresponding to this narrow range is further screened and then taken to more sophisticated models, which are in turn further validated.

Regardless of recent technological advances in drug discovery, the vast majority of lead compounds for treatment of cancer and infectious agents have come from natural products. Plant compounds thus provide the original source of novel structures, but not necessarily the final drug entity. In a scientific report published by Newman and Cragg in 2007, well over half of the sources for new drugs were naturally derived or semi-synthetic derivatives of natural products.

After several lead candidate compounds have been identified, they can be further screened for their behavior in a variety of tests. Ideally, lead compounds will be highly selective and biologically active toward the target molecule, are unlikely to interfere with other drugs, are water soluble, are able to easily penetrate the cell membrane (for easy delivery), and are free of intellectual property constraints.

When a compound has met these criteria, it enters what is referred to as the pre-clinical stage of development. At this point, Good Laboratory Practice data are collected to determine the plant-derived compound's safety profile and to learn whether the compound has sufficient merit to move forward to the clinical trial stage. In this stage, details such as how much of the compound is required to exhibit toxic effects both in cell culture and animal models is determined as well as how the compound works in the body and how the body metabolizes it. The sum of these data enables researchers to estimate a safe starting dose of the drug for initial human clinical trials. Good Laboratory Practice also assures regulatory authorities (such as the FDA) that the data submitted truly reflect results obtained during the study and can be reliably used when making risk/safety assessments.

When these points have been met, healthy volunteers will be enrolled in a small pilot study, using either the plant-derived compound itself or a modified, semi-synthetic version of it. These pilot studies enable researchers to gather further information so that they can design larger clinical trials, in which patients with predetermined characteristics will be recruited, the plant-based drug will be administered, and the patients' health will be observed and recorded over a defined period of time. The data collected will then be statistically analyzed.

Clinical trials are usually classified into four different phases and take several years to complete. Clinical trials can be expensive to perform and often are financed by a government organization or a pharmaceutical company. In a Phase I clinical trial, fewer than 100 healthy volunteers are recruited and the drug is assessed for its safety and mode of action in the human body. Phase I trials also include dose-ranging studies so that the appropriate dose for therapeutic use can be found. The tested range of doses will usually be a fraction of the dose that causes harm in animal testing. Phase I trials most often include healthy volunteers, whereas Phase II trials usually include populations of a few dozen to 300 patients and play a large role in determining the efficacy of a drug. This phase is critical for assessing whether a drug is too toxic to be considered further for use and whether it will work as predicted. Phase II trials are usually designed to be randomized and placebo-controlled, meaning that each human subject used in the study has been randomly assigned the drug in question or a placebo in its place. Phase II clinical trials are also designed to be double-blind, meaning that neither the subjects nor the researchers and their staff know which treatment is being provided to which patient. This prevents any bias in treatment, whether conscious or unconscious, from arising with the physician and staff who are administering the drug and caring for the patient.

If a Phase II trial is completed successfully, the drug in question enters a Phase III trial phase. Phase III trials include larger groups of 300–3,000 patients and often need to be conducted at multiple locations at once depending on which patients with the disease condition are available. This trial is completed to determine definitively how effective the drug is compared to the current "gold standard" treatment. Phase III trials can be both lengthy and expensive to run, especially for use as a therapy for patients who have chronic medical conditions.

Upon the successful completion of a Phase III trial, a drug can be submitted to a regulatory board for consideration in the commercial market. Once introduced to the general population, a Phase IV trial or post-marketing surveillance trial may be conducted. Phase IV trials can be used as a follow-up to gather further information about the drug, such as its potential to interact with other drugs, for example, or to detect any adverse effects over a longer period and for a greater number of patients than were present in the earlier clinical trials.

The impact of drug discovery and development has been to create a paradigm shift in traditional lines of thinking regarding the action of medicinal plants. Crude plant extracts with mysterious properties are now considered to be biological macromolecules that possess pharmacological attributes (Table 2.1). This shift in thinking was paramount for the identification of the chemical compound morphine, the active agent in the opium poppy *Papaver somniferum*, for example, or digoxin, a heart stimulant originating from the foxglove *Digitalis lanata*. Many modern drugs originate from plant sources and several examples are described in the next section.

Table 2.1. SELECTED EXAMPLES OF MEDICINAL DRUGS DERIVED
FROM PLANTS

Drug	Action	Plant Source
Acetyldigoxin, Digitalin, Digitoxin	Cardiotonic	Digitalis lanata (Grecian foxglove, woolly foxglove)
Betulinic acid	Anticancerous	Betula alba (common birch)
Caffeine	CNS stimulant	Camellia sinensis (tea, also coffee, cocoa and other plants)
Cocaine	Local anesthetic	Erythroxylum coca (coca plant)
Codeine	Analgesic, antitussive	Papaver somniferum (poppy)
Menthol	Rubefacient	Mentha species (mint)
Morphine	Analgesic	Papaver somniferum (poppy)
Quinine	Antimalarial, antipyretic	Cinchona ledgeriana (quinine tree)
Rhomitoxin	Antihypertensive, tranquilizer	Rhododendron molle (rhododendron)
Salicin	Analgesic	Salix alba (white willow)
Taxol	Antitumor agent	Taxus brevifolia (Pacific yew)
Teniposide	Antitumor agent	Podophyllum peltatum (mayapple)
a-Tetrahydrocannabinol (THC)	Antiemetic, decreases occular tension	Cannabis sativa (marijuana)
Valapotriates	Sedative	Valeriana officinalis (valerian)
Vincristine, Vinblastine	Antitumor, antileukemic agent	Catharanthus roseus (Madagascar periwinkle)

Source: Selected examples taken from About.com Chemistry Drugs from Plants Ethnobotany & Chemistry http://chemistry.about.com/library/weekly/aa061403a.htm.

EXAMPLES OF MEDICINES DERIVED FROM PLANTS

Taxol

A classic and well-known plant-derived compound originally identified as a result of bioprospecting is Taxol. Taxol is derived from the bark of the *Taxus brevifolia*, or Pacific yew, a very slow-growing conifer found in the old-growth forests of the Pacific Northwest (Figure 2.7). Commercially known as Paclitaxel, this drug is used to treat patients with various aggressively growing cancers. Taxol has a unique mechanism of action that has made it both a valuable tool for researchers and a critical agent in anticancer therapy. At the time of Taxol's discovery, only a small number of drugs had been identified that could slow down or arrest cell division. Compounds such as colchicine, for example, could be used to disrupt cell division by breaking

Figure 2.7.
Taxol, found in the bark of the Pacific yew tree, is a promising anticancer drug. Photo—National Institutes of Health, US Department of Health and Human Services, 2006.

down microtubules, molecules that make up the cell's scaffolding and maintain its structure. Cell division requires changes to be made in this scaffolding in such a way that microtubules have to be rapidly broken down and rebuilt. Unlike colchicine, Taxol prevents microtubule breakdown; the end result of this stabilizing effect is a reduced flexibility of the cellular cytoskeleton and ultimately a cessation of cell division. Since rapid and intense cell division is one of the fundamental characteristics associated with aggressive forms of cancer, Taxol can act to effectively target and block the spread of cancer cells. In addition, Taxol has been implicated as playing a role in inducing apoptosis, or programmed cell death in cancer cells.

The story of how a compound derived from the bark of the yew tree became one of the world's most important anticancer drugs is fascinating and is covered in detail in *The Story of Taxol: Nature and Politics in the Pursuit of an Anti-Cancer Drug*, by Jordan Goodman and Vivian Walsh. In the mid-1950s, the National Cancer Institute (NCI) was interested in searching for anticancer activities in a diverse assortment of synthetic and natural compounds. To this end, the NCI set up the Cancer Chemotherapy National Service Center (CCNSC), which set out to analyze samples submitted by external universities, institutes, and pharmaceutical companies. Part of the search involved a collaboration with US Department of Agriculture (USDA) botanists to screen for potential anticancer compounds in plants. In 1962, USDA botanists collected bark, twigs, leaves, and fruit samples from a Pacific yew tree growing in the Gifford Pinchot National Forest of Washington State. Crude extracts were prepared from these samples at the University of Wisconsin. The bark extract was found to have a cytotoxic activity on a sample cancer cell line as defined in a cellular assay in 1964. The active ingredient that displayed the anticancer effect, Taxol, was then isolated from yew bark a few years later.

During the 1970s, Taxol was taken through an array of procedures to determine whether it was a good candidate to be tested as an anticancer compound in human clinical trials. Out of 10,000 compounds tested at that time each year, only about five made it to an INDA (Investigational New Drug Application) filing with the Food and Drug Administration to initiate clinical trials using human subjects. In fact, 20 years later, the NCI had screened over 35,000 different plant and animal species for anticancer compounds. Due to the low number of new anticancer agents that were actually positively identified, the program was considered impractical and was eventually terminated in the early 1980s.

Although the identification of a new, potent anticancer drug from the bark of an indigenous North American tree was very exciting, a significant obstacle soon became apparent. To examine the compound in more detail for its therapeutic effects both in cell culture and in clinical trials, the NCI continued to require that larger and larger quantities of yew bark be collected and Taxol extracted and purified. For example, in 1969, almost 1,200 kg of bark was collected, which yielded only 10g of pure Taxol. The Pacific yew tree is very slow growing and the bark is extremely thin, so only a few pounds of bark can be harvested from each tree. Since 60,000 pounds of bark was predicted to be needed for just the Phase II clinical trials, it soon became clear that there would not be enough tree bark available to supply the demand for cancer treatment. The results of the Phase II trial were exciting; substantial improvements were demonstrated in patients with melanomas and refractory ovarian cancer. However, the NCI's Natural Product Branch estimated that the destruction of 360,000 trees annually would be necessary to synthesize enough Taxol to treat all the ovarian cancer and melanoma cases in the United States alone.

The identification of Taxol from the Pacific yew elevated environmental concerns regarding the bioprospecting of plants for medicines. There were other species of *taxus* that grew faster; however, the yields of Taxol that could be isolated from these trees were much smaller. In addition, the consistently highest yield of Taxol came from the bark itself rather from other parts of the tree, such as the needles, which unlike bark could be gathered routinely without impacting the health of the tree. Unfortunately, the yield of Taxol from yew needles was low and highly variable, making this form of extraction impractical.

Political tensions rose as it became clear that mass collections for Taxol were exacting a tremendous toll on old growth forests. Before the discovery of Taxol's activity, *Taxus brevifolia* was considered a "trash" tree, an understory tree with no extensive commercial use living on the banks of streams and in gorges and ravines. It had not been replanted on private forests as it was not valued as timber and was rapidly becoming depleted from the Pacific Northwest. Many of the *taxus* trees that still remained intact belonged to public reserves of old growth forest, part of complicated ecosystems involving both the trees themselves and the organisms surrounding them. Within the same time frame, the spotted owl, a resident of the same old growth forests as the Pacific yew, officially became an endangered species. As a result of this ruling, the Bureau of Land Management and Forest Service blocked the further harvesting of timber in spotted owl habitats.

Fortunately, Taxol can now be produced in quantity without damaging the environment further. In the late 1980s, a semi-synthetic production route to Taxol was developed at Florida State University. At present, a plant cell fermentation technology based on a specific *taxus* cell line has been developed for production of Taxol. This technology is much more efficient than semi-synthesis as the cells are propagated in large fermentation tanks from which the drug can be extracted and purified.

Aspirin

Another well-known drug that originates from a plant is aspirin, the most widely used medicine of all time. Aspirin contains the compound salicylic acid, derived from the bark of the willow tree. Unlike Taxol, salicylates were not stumbled on by use of the more modern bioprospecting techniques. Instead, salicylic acid has been used to relieve fever and pain since the beginning of recorded human history. The very earliest known reference to willow's medicinal properties comes from a Sumerian stone tablet, the Ur III tablet, from 3000 BC. The ancient Egyptians recorded the use of bark derived from the willow tree to treat various aches and pains, including stiffness and inflammation. The use of willow as a medicinal herb is listed in the Ebers Papyrus, a collection of medical knowledge that dates from the Middle Kingdom of Ancient Egypt, but more likely is a copy of an older manuscript written around the time of the Old Kingdom, dated 3000 BC, based on references cited within the text (Figure 2.8).

Willow bark remained a potent medicine and makes an appearance at the time of Hippocrates, and time and again throughout the history of medicine until 1758, when Edward Stone of England rediscovered and confirmed the ability of willow bark to reduce pain and fever. Stone

a
b

Figure 2.8.
(a) Page from Eber's Papyrus; (b) White willow.

stumbled across the effects of willow bark when he happened to taste it and found its bitterness reminded him of Peruvian bark, the source of quinine and imported from the New World to treat malaria. Fortunately, unlike the Pacific yew, the wide abundance and distribution of willow has not produced any environmental issues. The active ingredient, salicylic acid ("salicin" is from "salix," the Latin word for willow), was first isolated in 1828 by Joseph Buchner at Munich University. Unlike Taxol, salicylic acid is relatively easy to purify and synthesize in the lab. Eventually, in 1897, an acetylated form of salicylic acid, or ASA, which was easier on the stomach, was produced by the Bayer company in Germany. This medication was eventually patented by Bayer and given the brand name Aspirin.

In plants, salicylic acid is a phytohormone and plays a role in many physiological processes, such as plant growth and development, photosynthesis, ion uptake cell signaling, programmed cell death, and defense against pathogens. In humans, the mechanism of action is very interesting because salicylic acid has two general effects on the body. On the one hand, salicylates can inhibit the activity of cyclooxygenase-1 (COX-1), an enzyme that leads to the formation of prostaglandins that cause inflammation, swelling, pain, and fever. Salicylic acid also has an anti-platelet effect by inhibiting the production of thromboxane production. Under normal circumstances, thromboxane binds platelet molecules together to create a patch over damage of the walls within blood vessels. For this reason, ASA has also been prescribed at regular and low doses to assist in preventing heart attacks and strokes, and for patients who are at high risk for developing blood clots. As a result of aspirin's effect on stroke and cardiovascular disease, it is thought that aspirin may also play a beneficial role in preventing dementia and Alzheimer's disease, as well as pre-eclampsia, a complication in pregnancy, and possibly even cataracts and migraines. On the other hand, salicylates can prevent the production of prostaglandins that protect the stomach mucosa from damage by hydrochloric acid. Therefore, salicylic acid can provide both therapeutic action and potentially adverse side effects, such as increased risk of gastrointestinal bleeding.

Aspirin may be an effective treatment for certain types of cancer. For example, preliminary clinical studies have associated aspirin and other non-steroidal anti-inflammatory drug (NSAID) use with a reduced risk of colorectal and stomach cancer, prostate cancer, breast and ovarian cancer, and lung cancer. The explanation for this additional benefit of aspirin is based on the fact that some tumors overproduce a prostaglandin that assists in the cancer's growth and spread. Aspirin's ability to inhibit prostaglandin activity may therefore act by blocking the progression of a particular cancer.

Plant Compounds as Leads for Antimalarial Drugs

Malaria, caused by the protist *Plasmodium falciparum*, is a mosquito-borne tropical disease that infects over 350 million people a year (Figure 2.9). In rural areas of sub-Saharan Africa alone, where access to health services is often limited, malaria kills between 1 million and 3 million people a year, the majority of whom are young children. Mortality from malaria has in fact doubled over the last 20 years, largely due to the parasite's newly acquired resistance to antimalarial drugs.

Several antimalarial drugs have been available to treat malaria. One of the first effective treatments was quinine, and it remained the antimalarial drug of choice from the 17th century until the 1940s. Quinine is derived from the bark of the cinchona tree and originated as a traditional medicine of the Quechua Indians of Peru and Bolivia. Later, quinine was brought to Europe by the Jesuits, who ground the bark into a fine powder and mixed it with a liquid as an elixir to drink.

In the late 1940s, malaria was treated with the synthetic compound chloroquine. Chloroquine, which is relatively inexpensive and easy to produce, acts by binding to the heme of red blood cells and forming a complex that is

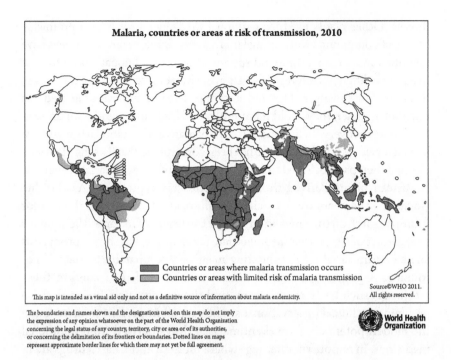

Malaria, countries or areas at risk of transmission, 2010

Countries or areas where malaria transmission occurs
Countries or areas with limited risk of malaria transmission

This map is intended as a visual aid only and not as a definitive source of information about malaria endemicity.

The boundaries and names shown and the designations used on this map do not imply
the expression of any opinion whatsoever on the part of the World Health Organization
concerning the legal status of any country, territory, city or area or of its authorities,
or concerning the delimitation of its frontiers or boundaries. Dotted lines on maps
represent approximate border lines for which there may not yet be full agreement.

World Health
Organization

Figure 2.9.
Malaria, countries at risk of transmission, 2010.

toxic to the parasitic cell. Unfortunately, the malaria parasite has developed widespread resistance to chloroquine, possibly due to the drug's overuse.

Another drug, artemisinin, derived from the plant *Artemisia annua*, has been used to treat drug-resistant strains of malaria. *Artemisia* has been used by traditional Chinese herbalists for over a thousand years. The active ingredient was identified in the 1960s in a major research program run by the Chinese army; it was the only traditional Chinese herb out of hundreds tested that effectively treated malaria. A disadvantage of artemisinin is that it does not rid the body of the malaria parasite completely. Symptoms reappear 3–4 weeks following treatment, most likely related to the short half-life of artemisinin in the body. In order to reduce the problem of recurrence of the disease, semi-synthetic derivatives of artemisinin with increased efficacy have been developed and are used in combination with other drugs to successfully treat malaria.

Regardless of these available medications, malaria remains a leading cause of morbidity and mortality in several African countries. Although artemisinin and its analogues have effectively treated chloroquine-resistant malaria, people in rural sub-Saharan African regions have no access to these medicines and/or cannot afford them. The need for new drugs that are inexpensive and easy to administer is now critical due to the widespread emergence of this disease. Other plants used in traditional medicines may offer a promising source of compounds with antimalarial activity. Ethnobotanical surveys have identified more than a thousand species of plants traditionally used to treat malaria by different cultures. The vast majority of these plants are not on any threatened or endangered list and since they have never been examined in a scientific manner, they could represent good "lead compounds" for antimalarial drug discovery. To this end, African shamans and traditional healers have been interviewed, and a database has been constructed that lists the plant species and their parts believed to contain elements with antimalarial properties.

African plants derived from plant genuses Cryptolepsis, Azadirachta (including the neem tree), Cochlospermum, and Guiera open the way for future research on the prevention and/or treatment of malaria. The aqueous root extract of *Cryptolepis sanguinolenta,* for example, has shown promise as an new antimalarial agent, exhibiting an efficacy comparable to that of chloroquine. A larger, randomized clinical trial is needed to confirm these results. As such trials are expensive and time-consuming, however, these preliminary clinical observational studies are all-important to drive home the hidden potential of novel antimalarial drugs derived from traditional medicines. In remote rural settings where modern antimalarial drugs are in scarce supply, preliminary research studies such as these can provide an initial evidence-based starting point that could be advanced further by both research institutions and philanthropic funding agencies. The fact that quinine and artemisinin, two of the most successful antimalarial drugs identified, are themselves plant-based and derived from indigenous knowledge, may convince those who are anxious to control drug-resistant malaria that looking for new lead compounds in plants is a good place to start.

HIV/AIDS and Traditional African Medicine

Nowhere in the world has AIDS (acquired immunodeficiency syndrome) had a more devastating effect than in the African continent, where tens of millions of people are confirmed to be positive for the human immunodeficiency virus (HIV). Traditional African medicines are also actively employed in the treatment of HIV/AIDS and for HIV-related problems. Two plant species in particular, African potato *(Hypoxis hemerocallidea)* and Sutherlandia, have been recommended by the Ministry of Health in South Africa and member states for use as immunostimulants in the treatment of HIV, despite the fact that traditional medicines like these have not been well researched and are poorly regulated (Figure 2.10). In some cases, these herbal remedies have been approved for use in conjunction with conventional antiretroviral treatments used in Western medicine.

a

b

Figure 2.10.
(a) *Sutherlandia frutescens*; (b) African potato.

African potato, with its bright yellow star-shaped flowers, has a long history of medicinal use on the African continent. This plant has been used for centuries by Zulu traditional healers for the treatment of a variety of ailments, from urinary infections to the common cold and even to arthritis and cancer. *Sutherlandia frutescens*, or the "cancer bush," is a flowering shrub with bitter, aromatic leaves with a strong reputation as a cure for cancer and a wide variety of other ailments. The aboriginal peoples of the Cape of South Africa, the Khoi San and Nama people, used *Sutherlandia* mainly in washing wounds and took it internally as a tea infusion to bring down fevers. The early colonists also regarded it as a successful medicine for the treatment of chicken pox, stomach problems, and internal cancers. It is still used today as a general cure-all.

As for actual biological activity, some indirect evidence has demonstrated that the roots of African potato contains sterols and sterolins, which may enhance the immune system in general. The means by which *Sutherlandia* may potentially act remains unclear, although initial reports have indicated that some antioxidant activity, immune stimulants, and anti-tumor effects may be present. While these preliminary findings are intriguing, there are many concerns that need to be addressed. For example, studies have demonstrated that the biologically active compounds of these herbal remedies may interact with HIV antiretroviral drug metabolism as a result of their inhibitory activity on efflux drug transporter systems. Therefore, patients being administered herbal medicines may be at risk for treatment failure, viral resistance, or drug toxicity.

Since little is known about how *Sutherlandia* interacts with other prescription drugs, researchers are now assessing its effects on cytochrome P450 enzymes, a family of enzymes that are involved in metabolizing prescription medicines, including HIV antiretroviral drugs. Alteration of P450 enzyme activity could have adverse effects on patients who are taking antiretroviral prescription medications. A clinical study is currently being conducted that compares P450 enzyme activity of patients before and after taking *Sutherlandia* with patients who take St. John's wort, a potent inducer of P450 enzymes. Another clinical trial is testing the safety and effectiveness of capsules of *Sutherlandia* in patients newly diagnosed with HIV.

Administering toxic plants for HIV treatment have also brought about severe adverse reactions, and sometimes even death. In Africa, cultural values play an integral role in the health care system and are often the primary form of care for treatment of people living with HIV/AIDS. On top of this, the accessibility of antiretroviral treatment for many Africans is extremely limited, due not only to its expense but also to the remote locations in which many people reside. For many Africans, there is a severe scarcity of physicians within their respective country (the ratio can be as poor as one physician to 1,700 people in some sub-Saharan African nations), whereas traditional healers are the trusted medical advisors who share a cultural identity with and regular access to the population (one healer on average per 200 people). A certain mistrust of the West in general can pervade African culture, resulting from memory of the treatment of native peoples during colonial days, a time in which shamans were often banned and sent to re-education camps. Today, healers, sharing the culture, can educate the population about AIDS even when such educational efforts are considered taboo. Several African nations have included traditional healers in educational campaigns to instruct the public on safe and hygienic practices, condom distribution, and other protective behavior.

Unfortunately, African traditional healers are thought by some to have played a role in the spread of HIV. A 3-year study was carried out to evaluate the role of traditional healers in the spread of HIV/AIDS in four continuous states in southeastern Nigeria, where HIV infection rates were known to be high. Twenty traditional healers and 69 patients were randomly selected and used in this study. Results showed that there was a serious HIV/AIDS-related risk inherent in the practices of the healers as indicated by the continuous usage of unsterilized instruments and cross contamination with patients' blood and body fluid in their practices. The study concluded that the practices of traditional healers themselves represented a mode of transmission of HIV infection in Africa.

Clearly, more effort is required to determine the safety, efficacy, and potential drug interactions of African potato and *Sutherlandia* as traditional medicines for the treatment of HIV. It is also necessary to collaborate with traditional healers to educate those providing alternative medical services against using unsafe practices. That said, the past policies of such entities as the Health Ministry in South Africa have been deeply disconcerting. For example, the previous health minister Manto Tshabalala-Msimang has been quoted as saying that she would not be hurried into providing antiretroviral drugs, and that beetroot and lemons were part of an ideal diet for people who were HIV positive. This is truly a terrible way of thinking for those in Africa living with HIV/AIDS. Today, instead of being handed a direct death sentence, North Americans who are HIV-positive are now able to live longer and more productive lives due to antiretroviral drug therapy. The same is clearly not so for the millions in Africa afflicted with this disease and who have no access to such drugs.

IMPACT OF BIOPIRACY, PRESERVATION OF BIODIVERSITY

What impact does bioprospecting have on the environment, in particular, on preserving biodiversity? Many experts in the field of environmental sciences have reacted with great alarm to the rapid extinction rate of a diverse number of organisms. It has been estimated that as many as 50 different species of plants and animals become extinct each day and that 1 million have been lost for good over the past 20 years. One reason that so many plants and animals are disappearing is due to the destruction of their habitats, as a result of deforestation practices stemming from rapid human population growth and uncontrolled economic development. There are still undiscovered species of organisms on earth, and whether they contain biologically active compounds that may benefit mankind is unknown.

Eighty percent of the world's people depend upon medicinal plants for their primary health care, and a significant proportion of the drugs produced are derived from plant-based leads. It is entirely feasible that the next "miracle" drug may at the moment reside within the genetic material of some endangered plant that grows in some tropical and poorly explored location, rich in biological diversity. Unfortunately, many habitats that display the most abundant biological diversity, such as tropical rain forests, happen to be in developing countries where their future is less stable due to fluctuations in economic and population growth. There is little incentive within these countries for cultivating sustainable development. These delicate ecosystems are constantly under stress, and developing countries often do not have the capacity to make their preservation a priority or the scientific know-how to implement it.

In Africa, a number of plants used in traditional medicine may soon become endangered. Besides the damage due to deforestation, other problems such as poor harvesting techniques for procurement of medicinal plant parts also impose a threat on species. For example, the bark of the *Prunus Africana* tree, used in traditional African medicine to treat many ailments, including prostate cancer, is often stripped in such a way that the tree is killed unnecessarily, leaving fewer appropriate specimens to harvest. If bark was taken from only one side of the tree, the tree would survive and produce more bark for future harvesting.

Maintaining biodiversity through conservation and preventing an increase in endangered plant species is critical for the continued pursuit of medical compounds from plants. Yet many locations where conservation is most precarious are also places that undergo uncontrolled and poorly planned economic development. In response to the seemingly opposing themes of environmental protection and economic development, the term "sustainable development" was born, that is, economic development that does not expend natural resources at a pace that would lead to the eventual loss of those resources. Sustainable development became the principal subject matter at the United Nations Conference on Environment and Development (UNCED), held in 1992 at Rio de Janeiro. The Rio conference resulted in the creation of the Convention on Biological Diversity (CBD), which aims to conserve biological diversity as well as to ensure equitable sharing of the benefits which come to pass from the use of genetic resources. Each country within the CBD is required to provide a national framework to protect and conserve the environment. The CBD also requires that a global consensus be reached regarding what constitutes conservation and depletion of biodiversity, and what steps must be taken to protect diverse species of organisms, their genetic material, and the ecosystems that support them.

INTELLECTUAL PROPERTY RIGHTS FOR INDIGENOUS PEOPLE

Indigenous people are all too often unaware of the value of their knowledge of native plant species. Many developing countries with vast biodiversity lack adequate laws to regulate the commercial use of compounds derived from plants. It is highly unlikely that bioprospectors would have achieved the same rate of success for identifying bioactive compounds in plants without the leads offered by indigenous peoples. It is difficult to measure how much indigenous knowledge constitutes intellectual property and how to protect this knowledge. Furthermore, it is an extremely difficult task to draw up a patent that adequately describes information derived from an indigenous population that has loosely inhabited a geographical region in small villages and has passed medicinal plant knowledge from one generation to the next through oral tradition.

As described previously, the Merck-INBio agreement appears to be one promising framework to build upon. Even this may be next to impossible to replicate in countries that at the moment lack the same business and governmental infrastructure as well as commitment to environmental conservation as does Costa Rica. For example, many undeveloped countries are run by corrupt governments that may not necessarily act in the best interests of their people or may in fact channel money meant for the holders of tribal knowledge for other purposes instead of distributing it fairly.

The exploitation of the rosy periwinkle stands out as an example of how some bioprospecting ventures have turned out for the worse. The identification and development of vinblastine and vincristine, chemical compounds found in rosy periwinkle to treat childhood lymphocytic leukemia and several other cancers, made hundreds of millions of dollars for the pharmaceutical company that developed the drugs but failed to provide any compensation for Madagascar, the country where the plant was found, or to the shamans who pointed out the plant in the first place.

Developing nations do not have the resources to analyze all of their biodiversity. Instead, developed nations will acquire patents for plant-derived compounds that are identified in developing countries. In some cases, once bioprospectors have searched for and identified an active biological compound from a particular plant species, often with the assistance of an indigenous tribal healer, they will search for other geographic locations where the plant is found. Then they cite as the place of discovery the region with the least legislation surrounding issues of profit sharing, thus preventing the indigenous people who led them to the correct plant species from any sharing of benefits.

While this attitude may be in a company's interest in the short term, serious problems arise from continuing to exploit indigenous knowledge without compensation. Indigenous information regarding medicinal plants that is kept and maintained by shamans and traditional healers may arguably be protectable as "trade secrets." Tribes and governments may no longer cooperate with pharmaceutical companies. In fact, Article 16 of the Convention on Biological Diversity has given developing nations this right.

For example, several nations in the South American Andes, including Colombia, Venezuela, Ecuador, Peru, and Bolivia, formed the Cartegena Agreement on access to genetic resources, which creates serious sanctions for companies that violate the agreement. Other countries such as Brazil have also created a network of legal sanctions to contend with possible biopirates. For example, Brazil has fined and imprisoned Dutch-born scientist Marc van Roosmalen on the grounds of biopiracy, regarding research he has conducted in the Amazon.

A highly cited example of the issues that arise pertaining to intellectual property rights and compensation of indigenous populations is that of the neem tree and its centuries-old use as an insecticide. The neem, or "divine" tree, has been used as a natural pesticide in Southeast Asia for untold generations and is described in more detail in the next chapter. Efforts to patent neem seed oil as an insecticide and fungicide were fiercely challenged by opponents' claims of biopiracy.

Issues such as unequal bargaining power between corporate entities and indigenous tribes can negatively impact the evenhandedness of an agreement. In some instances, the agreement appears to be fair, but in others it can more resemble an outdated colonial trade model involving the exchange of "trinkets for ivory." Agreements may also be drawn up that appear to be fair prior to the actual discovery/confirmation of the value of the plant-derived compound, a process that is by its inherent nature difficult to measure until much further down the research and development line. For example, even if a biological compound is identified from a plant that has been selected with the assistance of a shaman and is initially determined to potentially possess a useful medical application, it remains unclear at the time that the agreement is compiled exactly what the compound looks like, whether it can be replicated in the lab, whether it may be considered overly toxic in clinical trials, or whether it will require substantial modification before it can be commercialized. Will the compound turn out to be the next highly valuable anti-tumor agent or will it have a more modest use? Will it be turned down by the FDA or a similar agency in other countries as too dangerous or will it be too structurally similar to a compound derived from another plant species that has already been patented elsewhere? These uncertainties add a substantial level of risk (which may be likened to winning a lottery) for the pharmaceutical company and makes the actual worth of the plant-derived compound initially unclear at the time when the agreement is drawn up.

The matter of ownership of medicinal plants and indigenous knowledge presents other challenges. Some traditional healers feel that they are not adequately protected against some forms of biopiracy because their own countries lack laws recognizing their existence and knowledge. In fact, some countries in Africa still retain Witchcraft Acts, relics of past colonial times, which prevent any recognition of traditional medicine, thus negating any part it may play in the drug discovery process. To augment the process further, the same medicinal plants in question may be found far from their land of origin, residing in botanical gardens, herbariums, nurseries, and gene banks throughout Europe and North America, where they can be more easily accessed by bioprospectors, and unfortunately, by biopirates as well. Situations such as these give rise to the question of who owns medicinal plants in situ, and who owns the indigenous knowledge that is possessed by individual healers. Although many developing countries have signed the Convention on Biological Diversity, the regulatory framework to conserve biodiversity and implement intellectual property rights has not evolved sufficiently that it can be fully implemented by such nations.

CONCLUSIONS

Ethnobotany remains at the foundations of modern pharmacology. Originating from our past cultures, ethnobotany has existed over the life span of the human race through the combined knowledge of indigenous peoples past and present. This continuation of traditional medical knowledge is now being broken as indigenous cultures intermingle with the modern world and leave old customs behind. The potential to lose a great deal of ethnobotanical knowledge is imminent, particularly because much information is passed on from one generation to the next by the oral tradition and therefore has no permanent written backup record. The fact that only a small proportion of indigenous peoples have been interviewed at all for their knowledge of plant-based medicine confounds the situation even more. Only a small proportion of plant species have so far been tested, and the world's forests remain largely untapped as a source of uncharacterized bioactive compounds with potential applications to address global health problems.

As mentioned earlier in this chapter, natural products derived from plants play a dominant role in the discovery of leads for the development of drugs to treat human diseases. Until sialic acid was synthetically synthesized in the lab in the mid-1800s, virtually all medicinal compounds were plant-derived. Today, phytochemicals continue to provide a robust source of novel chemical structures to be used for drug discovery.

While searching for plant-based drugs for the benefit of humanity seems justifiable, many environmentalists oppose the pressure that bioprospecting places on the ecosystem of a given region and on sustainable development in general. Due to human activities such as logging, the demand for arable land for agricultural use, and increases in pollution, tremendous strain has been placed on some of the most fragile ecosystems on Earth. Moreover, weak regulatory frameworks, inadequate legal systems, and policy vacuums in general create situations that can be exploited either through biopiracy or through intellectual property legislation that monopolizes exclusive rights to natural resources, leaving indigenous people uncompensated for their knowledge and assistance. Clearly, maintaining the delicate balance between discovery of novel bioactive compounds in plants, the protection of the rights of indigenous peoples, and the preservation of the environment will be a significant challenge for years to come.

CHAPTER 3
The Lure of Herbal Medicine

"Herbal Medicines? Oh yes, I take them all the time to upkeep my health."

"Herbal Medicines? Ha! I can take you on a hospital tour of patients who have dabbled in herbal medicines. You will see how they ended up!"

Both of these comments, made by acquaintances, are representative of the wide range of attitudes toward herbal remedies that are found in North America today. The first statement comes from a shopkeeper, born in Asia and now residing in the United States. She claims to have maintained an herbal medicine regimen as part of her family life and culture as far back as she can remember. The second statement comes from a medical physician who works as a specialist in a Toronto hospital. This person is also Asian in origin, but in contrast, North American by birthright. She claims to see the dark side of the increasing popularity of herbal medicines in Western culture.

For those readers who are medical researchers and health practitioners: whether you like it or not, herbal medicines are taken by a strong majority of the world's population. As a result of the global marketplace and the immigration of peoples of diverse ethnic origins to the West, herbal medicines can be found virtually everywhere. Many people who are frustrated with the side effects of drugs prescribed for chronic health problems may turn to herbal medicine as an alternative. For others, herbal medicines represent a part of their culture that remains "tried and true." A number of this group of people may in fact live their entire lives in a Western industrialized country without having ever visited a board-certified physician.

The use of herbs to treat diseases is almost universal among nonindustrialized societies. Indeed, as mentioned in the last chapter, many of the pharmaceuticals currently available to physicians, such as opium, digitalis, and quinine, have had a long history of uses as herbal medicines. Many plants which provide active ingredients for prescription drugs came to the attention of researchers because of their use in traditional medicine. At least 7,000 medical compounds in the modern pharmacopoeia are derived from plants.

According to the World Health Organization, approximately 80% of the world's population presently uses herbal medicines in some form or another as a component of their primary health care. For most of these people, half of whom live on less than $2 US per day, pharmaceuticals are inaccessible and too expensive to be even considered as an option. Herbal medicines, in comparison, can be readily gathered and constitute little in cost.

While it is understandable that people who have next to no access to modern drugs continue to turn to herbal remedies, it is fascinating that so many Westerners also flock to these age-old practices. Some hold the belief that herbal medicines derived from plants are easier for the digestive system to tolerate and better for you in general. Obviously, not all plant products are good for you. Plants produce bioactive compounds for a reason—to deter pathogens, for example. These bioactive compounds are what give each plant species a competitive edge in its given environment— to avoid being eaten, to compete with other species, or to weather adverse conditions. It is coincidental that some of these plants can be helpful to human health. Even tomatoes, in their wild undomesticated form, contained a high degree of toxicity when eaten.

For anyone who has taken or would consider taking herbal medicines, a word of extreme caution is in order. Issues such as adverse side effects, inconsistent quality, adulteration with toxic contaminants, and potential interactions with other drugs are all very reasonable concerns and should be addressed wherever possible. Falling under the category of dietary supplements, herbal medicines are not regulated by the same governmental entities that handle other drugs, meaning that there is no quality control and the ingredients may in fact be different from what appears on the label. In many cases, the herbal medicine may not be efficacious at all, or often there are not enough data to determine whether it works. Other concerns such as using an endangered plant species or indirectly destroying its habitat also need to be considered.

Regardless, medicinal herbs have a long history of use in the practice of traditional medicines, and a substantial body of evidence has, over recent decades, demonstrated a range of important pharmacological properties (Table 3.1). Western biomedical researchers are examining not only the efficacy of the traditional herbal products but, through using a range of bioassays and analytical techniques, are developing improved methods to isolate and characterize their active components. Here we provide a brief overview of the history of herbal medicine from the beginning of civilization, with particular emphasis on Ayurvedic and Chinese cultures, where it still flourishes

Table 3.1. POPULAR HERBAL MEDICINES USED TODAY

Herb	Use	Efficacy	Safety
Echinacea	Upper respiratory tract infections	Inconclusive	No known side effects
Ginseng	Physical and cognitive performance	Inconclusive	Case reports of hyperactivity, restlessness
Gingko	Dementia	Likely effective	Case reports of bleeding
Garlic	Heart disease	Likely effective	Mild GI side effects, case reports of bleeding
St. John's Wort	Depression	Effective for mild depression	Numerous reports of drug interactions
Peppermint	Upset stomach	Inconclusive	Mild side effects
Ginger	Nausea	Inconclusive	No known side effects
Soy	High cholesterol	Effective	Long-term estrogen concerns
Chamomile	Insomnia/GI problems	No solid data	Rare allergic reactions
Kava kava	Anxiety	Likely effective	Case reports of liver failure

Source: Bent, S. Herbal medicine in the United States: review of efficacy, safety, and regulation: grand rounds at University of California, San Francisco Medical Center. *Journal of General Internal Medicine* 23(6) (2008): 854–59.

today. A number of examples of herbal remedies are provided, and how they fare under the scrutiny of modern science is detailed. The chapter ends with a discussion of how herbal medicine may in the future be combined with Western medicine to fight chronic diseases such as cancer.

HISTORY

The study of herbs for medicinal purposes can be traced back over 5,000 years of human history, to the Sumerians and later to the Egyptians. Evidence for the use of herbs in medicine has been recorded since the dawn of civilization. The Ebers' Papyrus, dating back to 1552 BC is the oldest preserved medical document and describes the diagnosis and treatment of a variety of illnesses using common herbal remedies. The ancient Greeks and Romans also used plants in their medical practices. Hippocrates, the famous physician of ancient Greece circa 460–370 BC, is sometimes referred to as the "father of modern medicine" and was known to use a few simple herbal drugs in his practice. Of great importance for herbalists and botanists alike is the *Enquiry into Plants,* a book that founded botany as a science and was written by Theophrastus (371–287 BC), Aristotle's successor at the Peripatetic School. Pedanius Dioscorides, a physician and botanist who practiced in Rome during the time of Nero, traveled across the ancient world, collecting medicinal herbs. In AD 40–90, Dioscorides compiled in Greek a five-volume *De Materia Medica*, a precursor to all modern pharmacopoeias and one of the most influential herbal books in history. Providing knowledge of the herbs and remedies used by Greek, Roman, and other cultures of antiquity, *De Materia Medica* referred to more than 500 plants and remained as a reference book until the 17th century (Figure 3.1).

Figure 3.1.
Page from Pedanius Dioscorides' *De Materia Medica*.

Little changed in the use of medicinal herbs during the Dark Ages of medieval Europe. Monasteries, which held in their possession copies of medical manuscripts from ancient Rome and Greece, were considered to be central places of knowledge and often cultivated herbal gardens for the treatment of general maladies (Figure 3.2). Herbal medicine was also practiced as a subdivision of general folk medicine, and a person skilled in healing played the combined role of physician, priest, prophet, and magician.

A number of plants were thought to possess healing powers that hinged on the supernatural. An example of this is mandrake, or *mandragora root*. The role of mandrake in herbal medicine and folklore is almost universal across cultures; its picture can be found in almost all medical and drug-related manuscripts across Europe (Figure 3.3). Indeed, the mandrake's fruit contains several plant compounds that have anesthetizing properties.

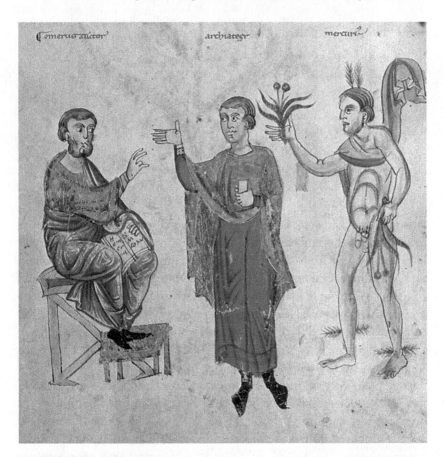

Figure 3.2.
Some herbs claim a mythical origin in history. Illustration of the poet Homer; on his knees lies a book with the inscription "Homerus" on the left side and the Greek "moly" on the right side. The physician stands the middle and on the right is Mercurius, carrying a Moly-plant in each hand. Moly was believed to be used as a remedy for pains in the uterus. From *De Materia Medica*.

Figure 3.3.
Mandrake (*Mandragora officinarum*), scanned from 15th-century manuscript *Tacuinum Sanitatis*.

However, the rootstock of the mandrake has historically been given the most attention. Legend has it that mandrake plants that grow under the gallows are derived from the sperm of the hanged. They are represented throughout medieval history as homunculi from the plant world. It should come as no surprise then that digging up these mysterious shapes was often accompanied by special ceremonies, as pulling mandrake plants up by the roots was determined to be a bad omen. In Shakespeare's *Romeo and Juliet*, for example, there is a reference to the blood-curdling scream that the root supposedly utters when it is pulled out of the earth. No one can forget the mandrake's cry in the current popular Harry Potter series. It is possible that since the mandrake root, with some imagination, can be thought to resemble a human form, it naturally became more important for medical practices than the fruit due to superstition rather than actual practical medical effectiveness. As magical as these roots may appear, their pharmacological effectiveness, if any, remains under dispute. Although some believe the root can be used as an aphrodisiac, there is no scientific justification to back up these claims.

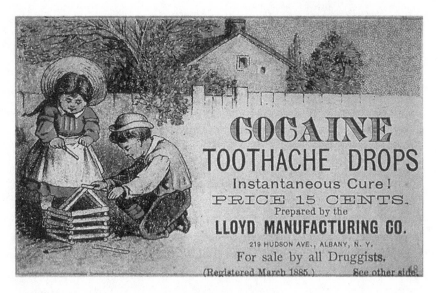

Figure 3.4.
Cocaine Toothache Drops, example of a patent medicine advertised in the United States, circa 1885. From the US National Library of Medicine, National Institutes of Health.

While the progress of medicine remained at a standstill in medieval Europe, herbal medicine developed and flourished in the Islamic world. Arabs were exemplary travelers and traders, and as a result they acquired plants and medical texts from the Far East and the West, allowing them to expand their knowledge. For example, in the ninth century, the physician al-Dinawari wrote the *Book of Plants*, which describes more than 600 plants actively used as drugs. In the 13th century, Ibn al-Baitar, a botanist/physician, produced the pharmacopoeia *Kitab al-Jami fi al-Adwiya al-Mufrada*; this book was consulted as a botanical authority until the early 1800s.

The 19th century brought about the emergence of patent medicines, concoctions containing questionable and often hazardous compounds. Patent medicines were often heralded as cure-alls for everything from epilepsy to tuberculosis. In many cases, herbs were described as the principal active ingredients, and their colorful names carrying an air of exoticism and mysticism were often maintained to further attract customers. All too often, the actual effect of these medicines came from the presence of narcotics or alcohols mixed in them, which might have relieved pain but also could cause drug addiction and even death (Figure 3.4). Eventually, with the impetus of works such as "The Great American Fraud" in *Collier's Weekly* by Samuel Hopkins Adams in 1905, the first Pure Food and Drug Act was set in place in 1906.

While patent medicines are now a thing of the past, a few vestiges remain. The soft drink, Coca-Cola, for example, is no longer sold for medicinal purposes, and its original ingredients, cocaine and caffeine, have been changed. The cocaine was derived from the coca leaf and the caffeine from the kola nut, leading to the name Coca-Cola. To this day, Coca-Cola uses as an ingredient a cocaine-free coca leaf extract. Kola nuts still act as a flavoring and a source of caffeine in Coca-Cola.

Another modern relic of the herbal exoticism that once characterized the patent medicine industry is the addition to shampoos of botanical compounds from herbs and fruit such as mangoes, bananas, and nectarines. No evidence exists that these herbal supplements enhance the health of one's hair, despite the attractive fragrances of these shampoos. Similarly, tablets or drinks composed of herbal mixtures, advertised as "nutritional supplements," are also commercially available and are marketed as remedies for various ailments.

THE BIOACTIVE COMPOUNDS IN HERBAL MEDICINAL PLANTS

Plants, like all organisms, consist of essential components or primary metabolites (sugars, proteins, and fats), which are necessary for growth and other fundamental processes of life, including reproduction. Secondary metabolites, on the other hand, are a class of molecules that help a particular organism survive and ensure that the plant can flourish within its environment. Consequently, a secondary metabolite is a more specialized compound, being found within one plant species but not another. The function of a secondary metabolite may be less clear but is most likely still paramount for the survival of the plant. For example, secondary metabolites such as pigments and pheromones can play roles in attracting pollinators, responding to environmental stresses such as drought or extreme temperatures, repelling insects or other predators, and even providing selective advantages over rival neighboring plants which are competing for the same light and soil. Plants have the ability to alter the synthesis of a particular secondary metabolite in response to a specific change in their environmental conditions, such as the presence of a pathogen or even a change in season. In this way, the profile of secondary metabolites produced by a plant will vary as the plant's environment changes. Examples of secondary metabolites found in plants include compounds such as alkaloids, phenols, and terpenoids. While chemicals produced as secondary metabolites serve a beneficial role for a particular plant species, they may also possess a therapeutic value or toxicity when applied to humans. Many plant metabolites, such as nicotine, function in plants as a powerful insecticide but are poisonous to humans when smoked or consumed. Other metabolites, such as the glycoside digitoxin of foxglove plants, are toxic to would-be predators yet possess medicinal benefits when utilized properly. Such metabolites have been refined to produce pharmaceutical drugs that are used in medicine today. Historically, therefore, the relationship between plants and medicine is both long and intimate. Indeed, the word "drug" itself is thought to be derived from the Dutch "druug," meaning "dried plant."

The use of plants for medicinal purposes has been intimately associated with the entire history of the human race. Although medicinal plants used in medieval Europe and ancient India and China are highlighted here, the utilization of plants in other persistent and enduring cultures including traditional African and Mayan tribes remains robust today. The following section provides a series of examples of medicinal plants that have been celebrated in earlier cultures, yet have survived the tests of time to become a continuous force in contemporary society. Here, a "Mythbusters" approach to herbal medicine has been applied by comparing the history and use of each medicinal plant in question with its performance when placed under the constrictions that exist for modern Westernized clinical trials. Conclusions about the effectiveness of each herbal remedy are drawn wherever possible.

St. John's Wort

St. John's wort, a highly popular herbal treatment for depression, has a history that is both long and colorful. In Ancient Greece, St. John's wort was used as a herbal remedy; during the Middle Ages, it was believed to keep demons at bay. The plant, officially known as *Hypericum perforatum*, received its more popular name based on the best time considered for harvesting the flower, St. John's Day, June 24 (Figure 3.5).

While the exact mechanism of action of St. John's wort is not entirely clear, it is believed to play a role in the inhibition of reuptake of the neurotransmitter serotonin, a principal mood regulator, in a manner that somewhat resembles the conventional selective serotonin reuptake inhibitor (SSRI) antidepressants. The major constituents believed to be responsible for antidepressant activity are hyperforin and hypericin.

A number of different human clinical trials have been conducted on the efficacy of St. John's wort and the results of some are promising. The Cochrane Collaboration, an international network that advises health care providers, policy makers, and patients, examined the results of 29 clinical trials consisting of more than 5,000 patients. From these trials, it was

Figure 3.5.
Photo of St. John's wort. Photo courtesy Saskatchewan Ministry of Agriculture.

concluded that extracts of St. John's wort were similar in efficacy to standard antidepressants and superior to a placebo used as a control in patients who have been diagnosed with major depression. In fact, the number of adverse side effects observed in patients who used St. John's wort was several times lower than some antidepressants available on today's market.

However, some of the studies which were included originated in German-speaking countries that have sometimes been criticized as reporting overly favorable results. A similar study performed by organizations affiliated with the National Institutes of Health (NIH) found that St. John's wort has minimal or no effects beyond the placebo in the treatment of major depression.

Unfortunately, St. John's wort interferes with the manner by which many drugs are processed in the body (Table 3.2). In particular, it affects the liver's cytochrome P450 enzyme system, which controls the transport of many drugs into the bloodstream. This results in variations in the level of any drug circulating in the blood that has been taken in addition to St. Johns wort and can lead to adverse reactions if the levels are increased or hinder the drug's effects if levels are decreased. The P450 system comprises at least 57 different genes in humans and encodes a series of membrane-associated proteins that are responsible for metabolizing thousands of substrates, including synthesis and breakdown of estrogen, testosterone, cholesterol, and vitamin D, as well as the end products of liver metabolism.

Table 3.2. DRUGS THAT POTENTIALLY INTERFERE WITH ST. JOHN'S WORT

Drug	Class	Effect of interaction with St. John's Wort
Alprazolam (Xanax)	Anti-anxiety	Decreased effectiveness
Aminolevulinic acid and other medications that increase sensitivity to sunlight	Photosensitizing drugs	Increased chance of sunburn, blistering, or rashes
Birth control pills	Contraceptive drugs	Decreased effectiveness
Cyclosporine (Neoral, Sandimmune)	Immunosuppressant	Decreased effectiveness
Irinotecan (Camptosar)	Anticancer drug	Decreased effectiveness
(Cytochrome P450 3A4 (CYP3A4) substrates)e.g., lovastatin (Mevacor), ketoconazole (Nizoral), itraconazole (Sporanox), fexofenadine (Allegra), triazolam (Halcion)(Cytochrome P450 2C19 (CYP2C19) substrates)e.g., amitriptyline (Elavil), carisoprodol (Soma), citalopram (Celexa), diazepam (Valium), lansoprazole (Prevacid), omeprazole (Prilosec), phenytoin (Dilantin), warfarin	Medications changed by the liver	Decreased effectiveness
Medications for depression,e.g., fluoxetine (Prozac), paroxetine (Paxil), sertraline (Zoloft), amitriptyline (Elavil), clomipramine (Anafranil), imipramine (Tofranil)	Antidepressant drugs	Decreased effectiveness
Medications for HIV/AIDS,e.g., nevirapine (Viramune), delavirdine (Rescriptor), and efavirenz (Sustiva).amprenavir (Agenerase), nelfinavir (Viracept), ritonavir (Norvir), and saquinavir (Fortovase, Invirase).	Protease inhibitorsNon-nucleoside reverse transcriptase inhibitors (NNRTIs)	Decreased effectiveness
Medications for pain,e.g., meperidine (Demerol), hydrocodone, morphine, oxycodone (OxyContin)	Narcotic drugs	Increases the effects and side effects of some medications for pain
tramadol (Ultram), sertraline (Zoloft), paroxetine (Paxil),nefazodone (Serzone),meperidine (Demerol)	Alters serotonin levels	Too much serotonin in brain leads to adverse side effects

Source: Taken from Natural Medicines Comprehensive Database Consumer Version. Natural Medicines Comprehensive Database, http://www.nlm.nih.gov/medlineplus/druginfo/natural/329.html.

Medications that may be affected by St. John's wort in this manner include anticancer, anti-HIV (human immunodeficiency virus), anti-inflammatory, antimicrobial, and cardiovascular drugs, which all utilize the same metabolic pathways. In addition, St. John's wort can interfere with certain medicinal herbs and supplements.

St. John's wort is not the only natural compound to affect P450 enzyme activity. Grapefruit juice, too, contains bioactive compounds that can inhibit the metabolism of certain medications and increase the risk of overdosing. As a result, removing grapefruit from the diet is strongly encouraged for anyone taking certain drug prescriptions.

St. John's wort has also been demonstrated to be poisonous to grazing livestock, such as cattle and horses. Animals that have consumed a large dose of this plant exhibit symptoms such as pawing at the ground, reduced appetite, increased heart rate, and elevated body temperature.

Garlic

A stellar example of a plant that has been used as both a food and a medicine all over the world for thousands of years is garlic, or *Allium sativum*. Acting as a food enhancement, garlic has also been shown to possess antimicrobial activities. Garlic is often referred to in old European folk tales as having the capability to ward off vampires, werewolves, and other evil supernatural creatures, possibly in reference to its perceived ability to protect against infections. In many cultures, garlic has been routinely consumed to prevent colds, hoarse throats, and coughs.

Many beneficial effects have been related to consuming garlic. It is believed that garlic's characteristic strong smell and taste are the result of sulfide compounds which are released upon cell breakage. While the pungent odor associated with garlic may play a role in protecting the plant against its natural pathogens, such as worms and insects, garlic also possesses a number of other chemical compounds that could contribute to a variety of potential health benefits. In a clinical setting, however, the reviews are a little more mixed. While garlic has indeed been shown to have antimicrobial activity in test tube experiments, these actions are less apparent in human clinical trials. Garlic has also been linked with a number of cardiovascular benefits, including lowering cholesterol and the risk of high blood pressure, and even for keeping certain cancers at bay. Since garlic has been shown to reduce platelet aggregation, it has often been recommended as an alternative to taking a daily recommended dose of aspirin.

How does garlic stand up in large clinical trials? Here the story changes somewhat (Table 3.3). Recently, a large randomized clinical trial funded

Table 3.3. LIST OF BENEFICIAL PROPERTIES ATTRIBUTED TO GARLIC

Thought to be effective for	Thought to be ineffective for
Preventing heart disease	Regulating blood sugar levels, preventing complications of diabetes
Preventing hardening of the arteries	Preventing high cholesterol
Preventing high blood pressure	Preventing breast, lung cancers
Reducing risk of stomach, rectal, colon cancer	Reducing ulcers
Antimicrobial activity	

Source: http://nccam.nih.gov/health/garlic/http://www.nlm.nih.gov/medlineplus/druginfo/natural/300.html.

by the National Institutes of Health concluded that garlic had little to no effect on reducing blood cholesterol levels. In fact, it can interfere with a number of drugs including warfarin and calcium channel blockers, as well as medications used to treat hypertension and HIV infection. Regardless, garlic, with its lively and animated past, remains a favored herb in many cuisines.

Ginkgo

The ginkgo tree has been identified in fossils dating back over 250 million years. Modern-day *Ginkgo biloba* grows best in environments that are well watered and drained. Ginkgo is chiefly considered to have memory- and concentration-enhancing qualities and has even been proposed to delay Alzheimer's disease and dementia. Clinical studies, however, provide contradictory results. While one study finds no evidence of memory enhancement upon intake of ginkgo, another finds a clearly positive effect. Ginkgo is believed to improve blood flow and prevent blood clotting. Side effects for those taking ginkgo include bleeding risks, digestive discomfort, headaches, dizziness, and heart palpitations. Ginkgo leaves can act as powerful allergens to some individuals. Other undesirable effects include adverse reactions associated with ginkgo co-administered with a number of drugs, including the anticoagulants ibuprofen, aspirin, or warfarin, and some antidepressants.

Valerian

Valerian (*Valeriana officinalis*), or all-heal, is a perennial flowering plant and was popularly used as a perfume in the 16th century. Applied as a medicinal herb by the ancient Greek physician Hippocrates, valerian is believed to act as a sedative and as a remedy for insomnia. The results of clinical trials regarding valerian's efficacy are at present inconclusive, or at best show valerian to be no more effective than a placebo. Furthermore, the precise mechanism of action of this herb in possibly causing drowsiness remains elusive. Nonetheless, few adverse events attributable to valerian have been reported. Since valerian in general depresses the central nervous system, it should not be used in conjunction with other depressants.

TRADITIONAL INDIAN MEDICINE

Ayurveda, the traditional system of medicine in India, has been a part of Indian culture for thousands of years and has its roots in both Vedic and Buddhist cultures. The term "Ayurveda" is a merging of the Sanskrit words *āyus*, meaning "life," and *veda*, meaning "knowledge" or "science." Ayurvedic medicine plays a significant and influential role in South Asian medicine today and is formally supported by the ministries of health in India and Sri Lanka. Ayurveda treatments include the use of herbs, meditation, massage, and yoga, and places an emphasis on prevention of illness and promotion of the balance of body, mind, and spirit. Hundreds of different plant species are used in Ayurvedic medicine as well as animal parts and minerals. For many, an intimate connection exists between Ayruvedic medicine and the theory and practices of a particular religion, such as Hinduism. A significant number of Ayurvedic physicians have claimed that their methods can also reduce chronic or stress-related conditions. In the West, a more modernized "New Age" version of Ayurveda has recently gained popularity as a unique form of complementary and alternative medicine (CAM). This Westernized, modernized version of Ayurvedic medicine has been reimported to India where it has played an important role in the country's tourist industry. For example, Kerala, a popular tourist area located in southern India, offers many recuperation and regeneration salons where the principles of Ayurvedic medicine are applied.

In India today, more than 100 colleges offer standardized degrees in traditional Ayurvedic medicine. In the United States, the National Institute of Ayurvedic Medicine, located in New York, carries out research based on Ayurvedic practices. The National Center for Complementary and Alternative Medicine (NCCAM) of the National Institutes of Health devotes a portion of its annual budget to research on Ayurvedic medicine, in an effort to determine how this form of alternative medicine stands in the context of Western medical science. As a traditional medicine, many Ayurveda products have not been tested in rigorous scientific studies and clinical trials; however, a few that have been tested show promising results.

Neem Tree (Azadirachta indica)

The neem tree, or "Divine Tree," as it is sometimes called, is native to India and Southeast Asia and has been used as both a medicine and a pesticide for many hundreds of years (Figure 3.6). Many parts of the neem tree, including the stem, leaves, roots, bark, gum, seeds, fruit, and flowers are actively used in traditional medicine. For example, oil, bark, and leaf extracts of the neem tree are used to treat a wide variety of ailments in Ayurvedic medicine ranging from leprosy to rheumatism, ulcers, and respiratory disorders; it is also used as a treatment for snake bites and even for malaria. In addition, extracts from the neem tree can be used as a natural insecticide and pesticide. Since neem products can become degraded by both ultraviolet light and rainfall, their relative levels of toxicity on the environment are low and thus they offer an ecologically safe alternative for pest control. For centuries, the leaves of the neem tree have been included along with dried turmeric in the storage of grain, as a natural insect and fungus repellent.

The neem tree has been the source of much scientific investigation in the search for novel biologically active compounds. To date, over 140 potentially

Figure 3.6.
Neem tree. Photo by J. M. Garg.

active compounds have been isolated from it; however, only a few of them have been examined in detail. For example, the compound nimbidin from seed oil of the neem tree has been shown to have anti-inflammatory, antipyretic, and antimicrobial activities. Nimbidin also has hypoglycemic activity and thus is a potential candidate for the treatment of diabetes.

Another compound known as azadirachtin is derived from neem seed and can act by preventing insect larvae from developing into adults. This compound has been shown to act as an antimalarial agent by preventing the growth of the parasite *Plasmodium falciparum*. The fact that azadirachtin even works in chloroquine-resistant strains of the parasite suggests that its mode of action differs from this commonly used drug. Neem oil included in kerosene lamps at low concentrations has been shown to act as a mosquito repellent. Neem oil is also used in sprays against fleas for cats and dogs.

A few preliminary clinical trials have been performed on neem products. For example, clinical trials using leaf extracts have shown that neem can indeed reduce hyperglycemia, suggesting a potential benefit for combating diabetes. Animal studies have also shown that garlic and neem used together exhibit a combined antioxidant and anticarcinogenic activity. As compelling as this sounds, high quantities of neem products, including neem oil and leaf and bark extracts, will cause adverse effects when used for medical purposes. Many components of neem such as nimbidol have exhibited toxic effects both on animal models and in cell cultures at sufficient concentrations. As a result, although the development of new drugs based on neem products shows great promise, more intensive research is required to ensure that the amount to be administered is safe.

Common Spices with Medicinal Effects

Some Indian spices that are very familiar to Western cultures also have been examined for potential health benefits. Here we look at five spices regularly used in food preparation: turmeric, cardamom, cinnamon, gingerroot, and black pepper.

Turmeric, a spice that comes from the root *Curcuma longa*, is a member of the ginger family and a native to South Asia (Figure 3.7a). In Ayurvedic practices, turmeric is thought to have several medicinal properties and many in South Asia use it as a readily available antiseptic for cuts, burns, and bruises. It is also used as an antibacterial agent and has been widely used by other cultures for the treatment of additional maladies. The scientific literature pertaining to the benefits of turmeric in medicine is extensive, to say the least. Over the last few years, for example, epidemiological studies have suggested that turmeric consumption may result in a reduced

a   b

Figure 3.7.
Illustrations from *Köhler's Medizinal-Pflanzen*. Author: Franz Eugen Köhler; (a) Turmeric (*Curcuma longa*); (b) Ginger (*Zingiber officinale*).

risk of some forms of cancer. Turmeric may also have other protective biological effects in humans, including anti-inflammatory, antibacterial, anti-angiogenic, and antioxidant effects. The result of extensive in vitro, animal, and human clinical studies have indicated that there are multiple biological effects of turmeric which act through several molecular mechanisms.

Cardamom, a spice also from the ginger family and grown in South Asia, is widely used to treat infections of the teeth and gums, sore throats, lung congestion, and digestive disorders, regardless of the fact that available scientific literature does not support any of these claims. The spice **cinnamon**, derived from the dried bark of *Cinnamomum verum*, is also used in traditional Indian medicine. Cinnamon has been used to treat diarrhea and other problems of the digestive system, and even to ward off colds. Several scientific studies have solidly demonstrated that cinnamon indeed contains antimicrobial properties. More recently, cinnamon has been shown to exert a protective effect on the liver.

Ginger, derived from the rhizomes of *Zingiber officinale Roscoe* (*Zingiberaceae*), has a long history of medicinal use for the treatment of a variety of human ailments including common colds, fever, sore throat, muscle aches, rheumatic disorders, gastrointestinal complications, motion sickness, diabetes, and cancer (Figure 3.7b). Ginger is on the Food and Drug

Administration's "generally recognized as safe" list, though it does interact with some medications, including warfarin. Tea brewed from ginger is a folk remedy for colds. In the same way, ginger ale and ginger beer are used for upset stomachs. Ginger contains several nonvolatile pungent phenolics including gingerols, shogaols, paradols, and zingerone, which may play a role in general beneficial health effects. Studies conducted in cell cultures as well as in experimental animals revealed that many of these pungent phenolics do in fact possess anticarcinogenic properties. Ginger acts as a strong antioxidant substance and may either mitigate or prevent the generation of free radicals. A number of mechanisms that may be involved in the chemo-preventative effects of ginger and its components have been reported; clinical data indicate that the aqueous ginger extract can lower blood pressure. Compounds within ginger such as zingerone are active against enterotoxigenic *Escherichia coli* -induced diarrhea, a form of diarrhea that is the major cause of infant mortality in developing countries. The anti-inflammatory properties of ginger have been known and valued for centuries. During the last two decades, a collection of studies have yielded strong scientific support for the long-held belief that ginger contains constituents with anti-inflammatory properties. For example, ginger's inhibitory effects on prostaglandin biosynthesis have been well documented. Furthermore, ginger has been shown to inhibit the induction of several genes involved in the inflammatory response, demonstrating that ginger modulates biochemical pathways activated in chronic inflammation. For scientists, this is exciting as it provides an opportunity for ginger products to become optimized and standardized concerning their effects on certain inflammation biomarkers.

Black pepper (*Piper nigrum*), a flowering vine from India, has also been used historically as both a food seasoning and for medicinal purposes. For thousands of years, the fruit, or peppercorn, of pepper has been used as a medicine for a variety of illnesses including gastrointestinal problems, heart disease, lung disease, insect bites, insomnia, joint pain, liver problems, toothache, and even eye problems. No medical evidence currently exists indicating that black pepper is effective in treating any of these ailments; however, it has been shown that piperine, a compound found in black pepper, increases the absorption of nutrients such as selenium, vitamin B, and β-carotene, as well as a number of drugs. While black pepper does not appear to be composed of the same level of helpful bioactive components as some other seasonings described in Ayurdevic medicine, it represents a good example of a medicinal herb that must be validated by modern biomedical science. Fifth-century writings of Ayurdevic medicine have documented black pepper as a treatment for eye problems, yet it is well known today that if applied as a salve or on a poultice as described in these texts, it would be irritating to the eye and possibly have damaging effects. Similarly, black pepper has been used from ancient times as a treatment for various stomach and intestinal problems; however, it has been shown to act as an irritant to the digestive system, and it is one food seasoning that medical practitioners often recommend avoiding after any form of gastrointestinal surgery.

Safety of Ayruvedic Herbal Medicines

In addition to the intrinsic toxicity of some Ayurvedic herbal medicines, one of the greatest safety concerns regarding the health risks of these medicines is their toxic heavy metal content. For example, one form of Ayrudevic medicine known as rasa shastra incorporates the addition of metals, minerals, or gems to the herbal medicine. This practice increases the likelihood of toxic metals contaminating the herbal medicinal product. A quick glance at the available scientific literature provides ample evidence that some Ayruvedic medicines are rife with toxic levels of heavy metals such as lead, arsenic, and mercury. For example, a 2004 study found the presence of toxic heavy metals in approximately 20% of Ayurvedic preparations made in South Asia and intended for sale in Boston. In addition to this, a study conducted in 2008 found that approximately 40% of rasa shastra medicines which could be purchased over the Internet possessed toxic metal contamination.

Serious illness and even fatalities due to lead poisoning have become associated often with traditional medicine use. A recent scientific paper was published which described the lead poisoning of a 60-year-old man with a history of diabetes and hypertension who consumed lead in the form of an Ayurvedic herbal remedy. The poisoning effect of this heavy metal fortuitously could be reversed by cessation of the herbal product, followed by lead-chelation therapy. As we increasingly become a more global community, it is imperative that those in the health care profession comprehend issues regarding safety and risk associated with traditional medical practices.

India has long been recognized as a major world source of herbs and spices, which serve not only as food seasonings but also are used medicinally. Efforts have recently been made toward determining whether these traditional uses of Indian medicinal plants hold up under scientific scrutiny. For example, a number of these Indian plants have been shown both clinically and in a laboratory setting to be a source of antioxidants, which are valued as playing a role in disease prevention. Based on some of the examples given here, spices such as turmeric, cinnamon, and ginger all have bioactive components and it may very well be to one's benefit to include them frequently in one's diet. However, the application of a spice generally used as a food seasoning for medicinal purposes can be confounding for several reasons: spices used in food are consumed in quantities that are not precisely measured; they are often combined with other spices which may mask other bioactive components; and eating itself is subject to highly socialized conditions. All of this creates difficulty in determining the relative benefits to one's health of Indian spices in the diet.

TRADITIONAL CHINESE MEDICINE (TCM)

Chinese herbs have been used for centuries. One of the first Chinese manuals that deals with medicinal herbs is known as the Shennong Ben Cao Jing (Shennong Emperor's Classic of Materia Medica) and dates back to the early Han Dynasty (Figure 3.8). Shennong, whose name translates as "The Divine Farmer," was a ruler of China some 5,000 years ago and is reported to have taught his people basic agricultural practices. Considered to be the founding father of traditional Chinese medicine, Shennong is credited with testing hundreds of herbs that have medicinal and poisonous qualities. Legend has it that Shennong had a transparent body and thus could see the effect of different plants and herbs on himself. In this way, Shennong could see which organ was affected and then select an antidote immediately if the effect was harmful. Also attributed to Shennong is the discovery of the plow, tea, and acupuncture. The earliest reference material regarding herbal medicine from ancient China, known as "The Divine Farmer's Herb-Root Classic," contains 365 medicines and is believed to be influenced by the teachings of Shennong. Ancient reference books such as the *Compendium of Materia Medica*, compiled during the Ming dynasty (in the 16th century), are still consulted today.

Plants play a paramount role in traditional Chinese medicine. Most Chinese medicines are composed of a cocktail of many herbs that is customized to a particular individual's needs. The Chinese medicine practitioner prepares a remedy using one or two main ingredients that are intended to treat the illness. Other ingredients are then added to create a specific formula intended to treat the patient's condition as a whole. On occasion, ingredients are added to negate the potential toxicity of a compound or to offset the adverse side effects of the medicine. Some Chinese herbal medicines require additional ingredients to act as a catalyst to improve efficacy. According to TCM philosophy, the balance of all ingredients and their interaction with each other collectively is thought to be of greater concern than the relative effect of the individual ingredients. A typical Chinese herbal medicine, therefore, can have only a few or as many as two dozen different ingredients, depending on the individual and his or her ailment.

Figure 3.8.
Shennong, the Farmer God, tasting herbs to discover their qualities. By Li Ung Bing, *Outline of Chinese History*, Shanghai, 1914.

The general philosophy of traditional Chinese medicine appears to be thousands of years old and stems from classic Confucianist, Taoist, and Buddhist beliefs that an individual's life and actions are intimately connected to the environment on many levels. According to this philosophy, the human body consists of a set of interconnected systems that must remain delicately balanced in order for good health to be maintained. TCM differs quite significantly from Western medicine by considering the physical body as a more general or "holistic" system rather than as specific anatomical parts. In order to prescribe the appropriate herbal formula and promote balance, a Chinese herbal medicine practitioner must determine the specific overall state of the individual. The body can thus be described in traditional Chinese medicine in a combination of various theories, such as the channel (meridian) theory, the wu-xing (five element) theory, the qi (vital energy) theory, and the yin and yang theory. Similarly, traditional Chinese herbs can be categorized into a series of systems such as the four natures, the five tastes, and the meridians. The four natures refers to the degree of yin and yang, namely cold (extreme yin), cool, warm, and hot (extreme yang). This is all taken into account when selecting the appropriate herbs for internal balance of a patient's yin and yang. For example, a medicinal herb that is considered to have a more "hot" yang nature would be considered if an individual complains of symptoms classified as "cold" or yin in nature. The five tastes of medicinal herbs are pungent, sweet, sour, bitter, and salty, with each of these tastes being capable of performing its own set of functions. The meridians relate to the specific organ channels the herb is reported to act upon. For example, menthol is believed to cool the lungs.

TCM Today

Traditional Chinese medicine has been part of Chinese culture and philosophy for thousands of years. In the 1950s, under Mao's rule, traditional Chinese medicine was modernized in the People's Republic of China. In fact, even today the Chinese government has encouraged the use of traditional herbs which have undergone analysis using Western medical standards as a cost-effective alternative to Western drugs. With a burgeoning population whose adherence to tradition renders it unlikely to give up TCM as a way of life, the decision of the Chinese Ministry of Public Health to modernize traditional Chinese medicine in alignment with Western standards is one practical way to keep medicine as affordable as possible. China, therefore, follows a strict set of rules for the utilization of medicinal herbs as drugs. Pre-clinical studies and clinical trials are required, just as they are for Western drugs. Chinese herbal medicines are now being standardized and expressed in terms of modern scientific indexes. The chemical structures of the active ingredients of hundreds of traditional Chinese medicinal herbs are now known, and in some cases, the ingredients can be synthesized in the lab (Figure 3.9). The therapeutic mechanisms of action of many traditional Chinese medicinal herbs have been elucidated and can be compared clinically with those of Western medicines. In fact, pharmacy schools in China now include the training of high-level personnel to complete the modernization of traditional Chinese pharmacy. These programs cover both the basic theories of traditional Chinese medicine and preparations of traditional Chinese materia medica as well as basic knowledge of modern science, including among other things, analytical chemistry.

Why Do Some Westerners Flock to TCM?

From a Western point of view, the recent fashionable trend depicting a "return to nature," to avoid potential toxicities and adverse effects of pharmaceuticals, has made Chinese herbology an attractive alternative to Western treatment of ailments as diverse as arthritis or cancer. Some

a

b

Figure 3.9.
Traditional herbal medicine in China today. (a) Institute of Plant Biopharmaceuticals and Traditional Chinese Medicine, Hangzhou Normal University, Hangzhou, China; (b) Chinese herbs for sale in shopping mall, Foshan, China. Photos taken by Kathleen Hefferon.

people feel that TCM provides another option to otherwise costly procedures which they cannot readily afford, or which are not covered by health insurance. In addition, the search for new bioactive compounds that may exist in Chinese herbal formulae for application in modern medicine has attracted the attention of many Western biomedical scientists.

For people of Asian origins who now reside in Western countries and cultures, the decision between choosing modern medicine and traditional Chinese medicine becomes even more difficult and convoluted. For many first-generation Chinese immigrants, TCM simply represents a system of health care with which they are most familiar and comfortable. The general philosophy of TCM tends toward a holistic approach to health, with emphasis on prevention of illness, while Westernized medicine is often characterized as being more reductionist, instead focusing on individual ailments.

The unification of spirituality and bodily health that is often embodied in traditional Chinese medicine makes prescribed treatments much more subjective than they generally are in Westernized medicine. For example, a visit to a TCM practitioner will often include an involved discussion regarding the overall physical, mental, and spiritual state of the client. The TCM practitioner will then incorporate this "big picture" of the client into the treatment plans. Such an overall diagnosis can often leave the client feeling more like an individual person rather than just a "number" as so often happens in a harried Western physician's overbooked schedule. As a result of this subjectivity, two independent TCM practitioners may treat the same individual with completely different herbal remedies.

First-generation immigrant Asians unfortunately can have language or communication barriers that make it quite difficult for them to discuss their illness with Western physicians. American residents may find that Westernized medicine is too expensive for those who have no or inadequate health care coverage. Simply booking and waiting for an appointment can be a long and involved process. To many who find the current health system in the West exasperating, treatment by a TCM practitioner would appear as a faster, more personable, and less expensive experience.

Of course, there are some people who prefer to experiment themselves with both TCM and Westernized medicine. An approach such as this can be potentially very dangerous due to the adverse effects of herb and drug interactions. Regardless, most Chinese in China do not regard traditional Chinese medicine and Western medicine as necessarily being in at odds with each other. While Western medicine is turned to in situations of medical emergency, belief in the use of Chinese medicine for preventive purposes remains strong. For those who prefer to sit on both sides of the fence, it certainly makes sense to visit the emergency wing of a Western hospital if they are having a heart attack, but they might also feel that the appropriate combination of herbs taken routinely might have prevented the heart attack from happening in the first place.

Examples of Traditional Chinese Medicines

Ginseng

Panax ginseng has played an important role in traditional Chinese medicine for thousands of years, primarily as a treatment for weakness and fatigue. *Panax ginseng,* whose name in Chinese can be translated as "man-root," refers to the forked shape of the root resembling the legs of a man. Ginseng herbal remedies are based upon the roots of several different plant species and have long been used for a variety of treatments, including to improve energy levels and increase resistance to stress; as a stimulant to increase physical and cognitive performance; to increase longevity; as an anti-inflammatory; to improve sexual function; and as a treatment for type II diabetes. Siberian ginseng (*Eleutherococcus senticosus*), American ginseng (*Panax quinquefolius*), and Asian ginseng (*Panax ginseng*) each has its own specific effects on the body. Dried roots are used in Chinese medicine. Ginseng is found in some energy drinks and teas that are quite popular in the West, although in these cases the concentrations of ginseng are in doses too low to be expected to have any beneficial effect.

It remains unclear as to whether ginseng has any of the earlier listed medicinal effects. Few clinical trials have been conducted under appropriate conditions, and many of those published present contradictory results. In many cases, the trials were poorly designed, or the quality and exact variety of ginseng used were unclear. The main active compounds in *Panax ginseng* are ginsenosides, which appear to simultaneously affect several biochemical pathways in the body, thus making it difficult to identify their true mechanism of action. Different ginsenosides may have different effects in pharmacology and mechanisms due to their different chemical structures. Furthermore, the level of each ginsenoside extracted from the plant can vary greatly depending on the plant species, where it is grown, and the time of harvest during its growth cycle. Various trials investigating the effects of *Panax ginseng* on several psychological parameters have ranged from positive effects to no effects at all. Similarly, clinical studies investigating the ability of *Panax ginseng* to enhance physical performance have produced no effect. However, ginseng has been shown in a number of studies to improve immune function as well as reduce symptoms due to diabetes. Furthermore, clinical trials have indicated that ginseng has been efficacious in improving sexual function and possibly exhibits anticancer effects.

While some mild adverse effects such as insomnia, restlessness and changes in blood pressure have been documented with the use of ginseng alone, adverse interactions have been identified between ginseng and certain drugs. For example, ginseng may interact with caffeine to cause hypertension. High dosage or prolonged use of ginseng may result in high blood pressure, acute asthma, and nose bleeds or excessive menstruation.

Ephedra (Ma-huang)

Ephedra, known in Chinese as Ma-huang and derived from plants of the genus Ephedra, has been used as a herbal remedy in traditional Chinese medicine for centuries to treat colds and other respiratory problems (Figure 3.10a). Active compounds include the alkaloids ephedrine and pseudoephedrine. Ephedra acts as a stimulant by constricting blood

a   b

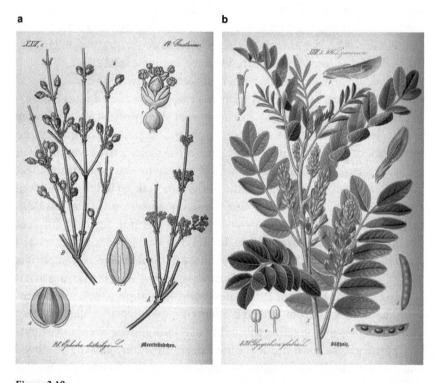

Figure 3.10.
Illustrations from *Flora von Deutschland, Österreich und der Schweiz* by Prof. Dr. Otto Wilhelm Thomé,1885, Gera, Germany. (**a**) Ma-huang (*Ephedra distachya*); (**b**) Licorice (*Glycyrrhiza glabra*).

vessels; it increases blood pressure and heart rate. Ephedra also increases metabolism and body temperature. Ephedra-containing dietary supplements have been widely used in the United States to promote weight reduction and energy enhancement. However, there are significant safety concerns regarding the use of ephedra-containing dietary supplements, especially when such use occurs by consumers without medical supervision. Ephedra-containing dietary supplements have been linked to adverse side effects including cardiovascular instability, and in 2004, the FDA banned the sale of ephedra-containing supplements.

Ephedra is included on the 2006 Prohibited List of the World Anti-Doping Agency. In the United States, the National Football League issued a ban on the use of ephedra as a dietary supplement for its players in 2001 after its involvement in the death of Minnesota Vikings offensive tackle Korey Stringer. In 2003, Steve Bechler, pitcher for the Baltimore Orioles, also died of ephedra-related causes. Ephedra remains widely used by athletes, despite the lack of solid evidence demonstrating that its use can enhance physical performance.

Ephedra has also been used as a weight loss supplement, and indeed may help some people lose weight over a short term. However, when taken in an unsupervised fashion, ephedra supplements have been demonstrated to cause a wide variety of adverse reactions, ranging from dermatological effects, delusions, heart problems, and on occasion, death.

Licorice (Gan-Cao)

Licorice comes from the root of the legume known as *Glycyrrhiza* and is found in some parts of Asia, as well as southern Europe (Figure 3.10b). In modern times, licorice extract is produced by boiling and condensing the root in water. Its main bioactive constituent is glycyrrhizin, used as a natural sweetener (it is 50 times sweeter than sucrose), and various types of flavonoids. As with other herbal medicines, the concentrations of each of these compounds may vary from batch to batch and thus affect the therapeutic effects. Licorice is also a popular herbal medicine derived from the dried roots and rhizomes of *Glycyrrhiza uralensis, G. glabra,* and *G. inflata* species. Powdered licorice root is an effective expectorant and has been used for this purpose since ancient times; modern cough syrups often include licorice extract as an ingredient. Licorice root is a traditional medicine used mainly for the treatment of peptic ulcers, hepatitis C infection, and pulmonary and skin diseases. Cankers or aphthous ulcers can also be treated with an herbal extract containing glycyrrhiza. CankerMelts GX patches, sold over the counter, contain a glycyrrhiza (licorice) extract. The use of CankerMelts has been shown to alter the course of the condition by reducing lesion size and may be as effective as amlexanox (which must be prescribed) in reducing pain and speeding up the healing process. Glycyrrhizic acid is now routinely used throughout Japan for the treatment and control of chronic viral hepatitis and latent Kaposi sarcoma. The pharmacokinetics of glycyrrhizin have been analyzed and its bioavailability has been shown to be reduced when it is eaten as licorice. Carbenoxolone, a synthetic derivative of glycyrrhizinic acid, has been used to treat sore throats. Carbenoxolone has also been used to treat oral lesions and type 2 diabetes; it is claimed to improve cognitive function in elderly men.

There are risks associated with consuming too much licorice. Licorice in large quantities can increase blood pressure, be toxic to the liver, and increase cortisol activity in the kidneys.

Chronic Illnesses and TCM

Traditional Chinese medicine is often considered as an alternative therapy for chronic disease, including some cancers. Chinese herbal medicines are believed to evoke a variety of cellular defense mechanisms, from apoptosis (programmed cell death) to inhibition of tumor cell division to enhancement of the immune system. For example, injections of Kaglaite (a substance derived from coix seeds) has been approved in China and used extensively for lung cancer, hepatic cancer, and gastric cancer. Since Phase I clinical trials have verified safety of the drug in the United States, the FDA has approved the initiation of a Phase II study for the treatment of lung cancer. Unfortunately, in general, the high degree of heterogeneity that exists among methodology, sample selection, and research design makes it difficult to assess the effects that many herbal medicines may have on cancer patients. An accurate database compiled from evidence-based clinical results simply does not exist.

The potential use of TCM as a treatment for irritable bowel syndrome (IBS) has also been examined. In a double-blind, randomized, placebo-controlled clinical trial, patients received either individually tailored treatments, a standardized Chinese herbal treatment, or a placebo for 16 weeks. The results of the study revealed that patients treated by TCM fared significantly better than the patients treated with placebos, and suggests that Chinese herbal medicine may offer an alternative to current pharmaceutical approaches for patients with IBS.

Since it appears that at least some herbal products do in fact possess beneficial medicinal qualities, why not, wherever possible, replace conventional, Westernized drugs with less expensive natural products derived from herbs? In the United States, a primary constraint would be the inherent problems involved in regulating herbal medicines. Under current US law, medicinal herbs fall under the category of dietary supplements. While the FDA may remove a product from sale should it prove to be harmful, manufacturers maintain the right to produce, sell, and market herbs without first demonstrating safety and efficacy, as is required for all pharmaceutical drugs marketed in this country. In the public perception, herbal products are still largely considered to be safe; perhaps some of this belief has been drawn from memories in the not so distant past before the pharmaceutical industry became so prominent and medical practitioners routinely relied on plants and their related products for treatment. To address the use of traditional medicines, in 2002 the Medicines Control Agency published the report "Safety of Herbal Medicinal Products." The primary issues of concern revolve around quality control, presence of contaminants, adverse effects or toxicity, interactions with other drugs, efficacy in a clinical trial setting, and even endangerment of a plant species. The following section expands on some of these concerns.

Quality Control, Lack of Standardization

Of principal concern is the lack of standardization and resulting poor quality control that is found for herbal medicines. There is no guarantee that a consistent dosage of a particular active ingredient will exist from one plant to another, even if one is weighing the same amount of roots or leaves consistently. The quantity of an active ingredient will vary depending on the growth conditions of the plant (how much sunshine, moisture, etc., was available), the location of the plant (on a hillside in India, a plain in Asia, or perhaps from fields within the southeastern United States). Even if the herbal medicine has been processed into a pill or tablet, for example, these are often not standardized, nor is the shelf life of the active ingredient taken into account. No quality control is implemented to prevent inconsistencies in content from one pill to the next, or from one company to the next. In fact, there is no guarantee that the botanical identity of the plant species claimed to be sold in the package is even correct (as has been shown in many published instances which I found with a little investigation). Unfortunately, herbal medicines that are adulterated with plants of a similar-looking species are not uncommon. A process therefore is lacking that can provide assurance to the customer that the plant listed is indeed found in the package, and that plant growth, harvesting, storage, and processing all follow good agricultural practice and good manufacturing practice guidelines.

The issue of standardization and quality assurance has become a matter of great interest to the Chinese with respect to TCM. Analysis of well-characterized marker compounds is one of the most popular methods for identifying the herbal materials and for quality control of TCM—for example, ginsenoside profiling to authenticate that a herb is actually derived from *Panax ginseng*. Advances in molecular biology in the past few decades have created genetic tools that can now provide more standardized and reliable methods for authentication of herbal materials at the DNA level.

Toxins and Contaminants Present in Herbal Medicines

In addition to problems of quality assurance that abound for herbal medicines, many safety concerns exist regarding adverse reactions to herbal medicines, including the presence of toxins and contaminants in herbal remedies, allergic reactions, and herb-drug interactions. As mentioned earlier, one study published by NCCAM in 2008 indicated that approximately one-fifth of the Ayurvedic medicines tested were found to contain detectable levels of lead (most common), mercury, or arsenic. In fact, all the metal-containing herbal products that were identified exceeded the standards for what is considered to be an acceptable daily metal intake. In the United Kingdom and Canada as well, a Chinese herb known as *Fufang Luhui Jiaonang* was taken off the shelves upon discovery that it contained approximately 11% mercury. Besides heavy metals, other potentially toxic compounds have been added to Chinese medicinal mixtures, following the concept that a poison should be used to cure a poison. Furthermore, as in the 2008 Chinese milk scandal, some herbal mixtures have been found on occasion to be laced with additional chemicals that may alter the intended effect of a herbal preparation or prescription.

Adverse effects that are directly due to the active ingredient within the herbal mixture can also occur. As mentioned previously in this chapter, the Chinese herb ephedra was banned by the FDA in 2004. This ban was to target the use of ephedra in Westernized weight loss products, which had led to a number of adverse effects, including death in some instances. Traditional Asian preparations of ephedra were excluded from the ban, as there were no cases of ephedra-based fatalities with patients using traditional Asian preparations of the herb for its traditionally intended use of treatment of respiratory ailments.

Another example of the danger of some herbal remedies can be found in the kava plant. Although scientific studies provide some evidence that kava may be beneficial for the management of anxiety, the FDA has issued a warning that using kava supplements has been linked to a risk of severe liver damage. Adverse side effects of other herbal medicines include cardiovascular toxic effects, hematologic toxic effects, neurotoxic effects, nephrotoxic effects, carcinogenic effects, and even allergic reactions. Herbs as benign as chamomile and echinacea, both of which belong to the daisy family, a group of related plants which include common allergens such as ragweed, can have an allergenic effect on some individuals. It is entirely possible that administration of these herbs could result in an allergic reaction including skin rashes, increased asthma, and even anaphylaxis.

Furthermore, many herbal medicines have been demonstrated to interact in adverse ways with prescription and over-the-counter pharmaceuticals that have been assigned to the patient. For example, St. John's wort has been demonstrated in many experiments to interact with a number of drugs by affecting the way the body processes or breaks down the drug. Drugs that can be affected include antidepressants, birth control pills, cyclosporine, digoxin, warfarin and related anticoagulants, and some drugs used to treat cancer and control HIV infection. When some herbal remedies which have the effect of lowering blood pressure are used in combination with a medication prescribed for the same purpose, one's health can be endangered. A large number of patients take herbal medicine for any number of reasons, and they do so without informing their physicians. It is imperative, therefore, for patients to disclose to a physician any herbal medicine that they may be taking, as well as any that they plan to take or stop taking. In a recent report from a research group in Aberdeen, Scotland, for example, a healthy 25-year-old man who was admitted to hospital with a difficult-to-control nosebleed was found, upon further questioning, to have recently started self-medicating with a combination of garlic tablets, aspirin, and milk thistle. All three may have contributed to the difficulty in management of the nosebleed. This case highlights the importance of asking about complementary and alternative medicines when taking a patient's history, as their use may have implications for management.

Efficacy in Clinical Trials

One of the most significant drawbacks of herbal medicines is the general lack of information regarding their relative safety and efficacy, meaning how well the herb safely performs its function, or whether it can perform its function at all. Clinical trials, or research studies in which the safety and efficacy of treatments and therapies are tested in people or animal models, are essential for determining which treatments work, which do not, and why. Designing and carrying out a clinical trial is both expensive and difficult. A large number of volunteer patients must be carefully screened and selected. Clinical trials are often conducted in a random, double-blind fashion, meaning that neither the individual patient nor the scientist who conducts the trial knows whether the drug (in this case, herb) or a placebo is being administered. A key describing which sample was provided to which patient is maintained in confidence by a third party, and this information is not released until the conclusion of the trial. All controls, both positive and negative, must be carefully considered and included. It may take several years to run through a single clinical trial. The data collected from the trial are then subjected to intense epidemiological analysis. To run a new active compound through this manner of intensive testing requires a great deal of time, cost, and expertise.

Some herbs do indeed exhibit positive results in in vitro studies as well as animal models or small-scale pilot clinical trials, and a number of the herbs which have been demonstrated to be efficacious show great potential. However, more than a few studies on herbal treatments have produced negative results. For example, in a published study of hundreds of Chinese herbal medicines tested as potential candidates claimed to be used for treating malaria, only one was found to actually be effective. This particular medicine, known as artemisinin, was isolated from Chinese wormwood (*qinghao*) and is now used worldwide to treat multidrug-resistant strains of malaria. Artemisinin is also under investigation as an anticancer agent.

Moreover, even for published clinical trials using herbal remedies that demonstrate positive activity, the quality of the science has often been questioned. Too often, problems such as a lack of appropriate controls, a reporting bias, and a lack of rigor in the design of the clinical trial lessen the credibility of the study. In many cases, the cumulative evidence from several clinical trials is still not fully convincing because the design of the trial is questionable and the quality too poor for the results to be trusted. In addition, the molecular mechanisms and pharmacological effects often remain unclear, making the herbal treatment less acceptable to conventional medical practitioners.

Endangered Plants, Property Rights, Patent Issues

Other problems exist regarding the ownership of the herb and/or active compound that individuals as well as governments may need to address. Patent issues and proprietary licensing are primary examples, as are protection of endangered plant species. In 2008, the Botanic Gardens Conservation International stated that hundreds of medicinal plants are at high risk of extinction. As the demand for botanical products increases, new issues must be addressed. For example, picroliv, derived from the endangered medicinal plant *Picrorhiza kurroa (kutki)*, constitutes an important component of many Indian herbal preparations, used mainly for the treatment of liver ailments. Picroliv is in high demand in both national and international markets due to its apparent ability as a strong hepato-protective and immune-modulatory compound. Similarly, many traditional Chinese medicines often incorporate ingredients from animals (for example, seahorses, rhinoceros horns, and tiger bones), some of which are endangered. The use of endangered plant and animal species has resulted in a black market of banned animal and plant parts. I was very sad to hear upon a recent trip to China, for example, that although they are quite illegal, I could likely buy many of these myself, if I was willing to come up with the right price.

WHAT SCIENTISTS NEED TO KNOW ABOUT HERBAL MEDICINES

Herbal medicines are often dismissed or even belittled by scientists who investigate purported claims made by herbal practitioners when these are found to be inaccurate or even untrue. While circumstances such as these do indeed take place, there are some points to be brought forward regarding the negative conclusions of these scientific, evidence-based assessments.

The Intrinsic Differences Uncovered in in Vitro versus in Vivo Tests

The term "in vitro test" refers to experiments that are performed in the absence of living tissues or animals—that is, experiments that may be performed entirely in a test tube. Initial pharmacological tests are often performed in vitro first to determine the biological activity of a particular compound derived from a plant or a crude herbal extract. Such small-scale tests are less expensive and require a smaller amount of material. The lack of need for test animals circumvents any ethical concerns which may arise. For example, as discussed in the bioprospecting chapter, some of the earliest in vitro bioassays were used to identify plant compounds that exhibited antimicrobial activity; these would be followed by tests to determine cytotoxicity using mammalian cells, and so on.

It is not always easy to isolate a single active compound from a complex, crude plant extract. One would expect that the biological activity of a herbal remedy is likely due to the actions of an individual compound within the extract. Using this line of thought, it would seem reasonable that by fractionating an herbal extract into its individual constituents and testing each of these for its relative biological activity, one should be able to identify the compound in question. While this "reductionist" approach does indeed work in many cases, sometimes it does not. In some cases, for example, the activity is lost or greatly reduced upon fractionation of the original plant extract; results such as these imply that the purification techniques used may have damaged the active compound. Alternatively, perhaps two or more compounds are required for biological activity, and once they are separated from each other by fractionation, they lose their ability to produce an effect entirely. Indeed, many herbalists claim that the healing effects of their medicines are the result of an appropriate mixture of ingredients rather than the individual actions of one or two compounds present in the plant extract. On the other hand, in some cases, several compounds present in the plant extract have the ability to exert the same general biological effect but perhaps at different levels of intensity. Such circumstances make it even more difficult to isolate and identify the active component of an herbal remedy.

Even if a biological activity is identified for a particular herbal extract using in vitro methods, it can often be difficult to extrapolate from these results an appropriate dose to ensure observable activity in an animal or human being. Furthermore, the means by which the plant is administered to an individual may have profound effects on its ability to act successfully. If one is to ingest the herbal remedy orally, for example, issues such as a person's ability to digest the plant in the gut, absorb constituents into the bloodstream, and metabolize the active ingredient in the liver may bring about significant reductions in efficacy. In addition, other plant components may have a positive or negative effect on absorption and metabolism of the active ingredient, thus changing its pharmacokinetic profile to one that differs substantially from the original, in vitro tests.

In many cases, the chronic disease state or illness is much more sophisticated and complex than the in vitro test used to first identify the biologically active plant compound; therefore, it is not realistic to expect that the results of the in vitro and in vivo tests can be at all comparable. For example, plant extracts which are proposed to potentially possess anticancer effects may be tested under many different circumstances, such as with several different cancerous cell lines, in an attempt to identify an appropriate in vivo environment upon which the extract may work. Situations such as these demonstrate how difficult it can be to evaluate a proposed herbal remedy for its beneficial properties in a clinical setting.

The criteria by which experimental research in general examines any new drug for its potential use in medicine has both strengths and weaknesses. In the case of herbal medicines, it is not beyond one's imagination to see that clearly in some instances, these criteria may not be adequate to address fully whether the remedy has some potential. The fact that many clinical trials that are poorly designed provide a plethora of conflicting results does nothing to shed light on the picture in general. In the future, compounds identified in plants as having potential medicinal benefit will require scientific analysis that takes these concerns into account.

COULD HERBAL MEDICINE AND WESTERN MEDICINE COMPLEMENT EACH OTHER?

In many ways, herbal and modern Western medicinal approaches and philosophies appear to represent opposite ends of the spectrum. While many plant secondary metabolites are biologically active compounds that do indeed possess medicinal value, most people who describe themselves as herbalists do not use particular isolated phytochemicals for their remedies; rather, they tend to use extracts from plant roots or leaves, containing many components. Pharmacologists prefer to use single active ingredients so that the dosage can be quantified more easily. Herbalists, on the other hand, disapprove of the concept of a single active ingredient. They believe that the combination of bioactive compounds present in a herbal remedy will strengthen its therapeutic effect overall and reduce any potential toxic effects. They offer the argument that multiple health effects can be attributed to a single active ingredient. Herbalists believe that similar remedies cannot be replicated in a laboratory setting, rather, they propose that herbal remedies affect each individual patient in a unique manner. Pharmaceutical researchers also recognize the importance of drug synergism but would prefer that the ingredients used in a herbal remedy remain consistent, and that its efficacy be tested under the rigors of a clinical trial.

It is currently estimated that more than half of all patients diagnosed with cancer explore complementary and alternative medicine, with particular emphasis on herbal medicine. A number of comprehensive reviews have been written in an attempt to assess the safety and efficacy of herbal medicines commonly used by patients to treat cancer. Ginseng, a scavenger of free radicals, has shown promising clinical results as a protector of DNA damage due to radiation therapy. Ginseng could be used as a

follow-up treatment for cancer patients to prevent adverse effects resulting from radiation treatment. Are efforts being made to explore herbal
and other forms and practices of alternative medicine as potential and safe
complements to Western medical practices? Absolutely. Today, the practice of preventive medicine and the role of the patient in self-management
of care are being emphasized more than ever and are reflected in many
government health care policies around the world. As a result, herbal
medicine is undergoing both more attention and more scrutiny. In the
United States, for example, the National Center for Complementary and
Alternative Medicine or NCCAM, in the National Institutes of Health,
places these alternative healing practices in the context of rigorous science, including the funding for clinical trials of the effects of herbal medicines, and releases information on the efficacy, adverse effects, and
potential herb-drug interaction to both the public and health profes-
sionals. In Europe and the United Kingdom, the Traditional Herbal
Medicinal Products Directive on the licensing of herbal medicines was
introduced in 2004. This Directive was set in place to respond to increasing public interest in the use of herbal products and to control herbal
products and herbal practitioners. Within the European Union (EU),
herbal medicines that demonstrate efficacy in clinical trials acquire product licenses. In order to accommodate the increase in use of herbal medications, the EU Traditional Use Directive has recommended that evidence
of 30 years' use, of which 15 years must be within the European Community, would be sufficient to consider the herb safe.

In China, a degree of integration between traditional Chinese and
Western conventional medicine also exists. In some instances, TCM
has been integrated with conventional medicine to combat cancer. For
instance, at any particular hospital in China, a patient may be seen by a
multidisciplinary team and be treated concurrently with surgery, modern
drugs, and a traditional herbal formula.

Due to the efforts made to integrate traditional Chinese and Western
systems of medicine, the number of databases available recording compi-
lations of herbs, herbal formulations, phytochemical constituents, and
molecular targets is increasing. Part of this is due to the widespread popularity of TCM and herbal remedies from other parts of the globe and their
potential interactions with Western drugs. Using cancer treatment once
again as an example, a combination of herbal therapy and chemotherapy
is under evaluation. In an ongoing clinical study to evaluate the effectiveness of TCM in reducing the relapse and metastasis of stage II and III
colorectal cancer, 202 patients who were diagnosed as suffering from this
disease were routinely treated with Western medicinal techniques,
including surgery, or chemotherapy or/and radiotherapy. Patients were

placed into two groups based upon whether they were also treated with TCM during remedial therapy and their progress recorded for 1 to 5 years. So far, the study has indicated that patients who took the herbal remedy as part of their remedial treatment took much longer (26.5 months, on average) to experience a relapse of the cancer than patients who did not receive herbal medicine (16 months, on average). The results of this study indicate that combining TCM and Western medicine may offer significant clinical value and a real potential for decreasing the relapse or metastasis rate in stage II and III colorectal cancer.

CONCLUSIONS

Whether one has a scientific background or not, the field of herbal medicine appears to be highly complex and controversial. Of great fascination are the connections between philosophy and spirituality that underlie many aspects of traditional herbal medicine, regardless of the specific culture in which it is rooted. Repeated themes of holism and an intimate connection of the body with nature resonate throughout Ayurvedic, Chinese, and other ancient cultures. Many of these civilizations, including Hippocrates's Greece from which modern Western medicine originates, also described the body using pre-empirical terms of underlying forces.

For some herbal medicines that have been maintained over many generations in the form of a written text, intrinsic contradictions in the ancient scripts themselves can provide a great challenge with respect to maintaining credibility of their actual health benefits. The employment of some plants in modern medical practices have a long history of use for the same ailments. The traditional Chinese remedy of using menthol to treat sore throats, for example, has been corroborated with Western medicine. Some plants, however, appear to have no connection whatsoever and in fact be quite the inappropriate remedy. The Ayurvedic use of black pepper in a poultice to treat eye problems, for example, is a treatment that no one today would recommend, and thus seems rather counterintuitive. One is left with the impression that while some remedies are on target, others are wide of the mark.

Often consumers dangerously believe that since herbal medicines are natural, they should be safe to use. On many occasions, herbal medicines which lack any proven efficacy have been used to replace conventional medicines that have a solid and robust track record of proven efficacy. This misleading oversimplification neglects many safety issues such as poor governmental regulation, potential toxicity due to the herbal medicine itself or due to other contaminants associated with the medicine, and possible interactions of the herb with other prescription drugs. In order to fill the gaps in knowledge regarding the safety and efficacy of herbal medicines, better record keeping and more information in general are required. Without question, well-designed double-blind clinical trials are needed to determine the safety and efficacy of each plant before it can be recommended for medical use. Unfortunately, only a few herbal remedies have been tested this closely. In many cases, there is simply not enough information to provide appropriate advice to patients.

Herbal medicines are used widely in the United States, and according to a recent survey, the majority of people who use herbal medicines do not inform their physicians about their use. Herbal medicines can cause abnormal test results and confusion in proper diagnosis. Herbal medicines can alter test results by direct interference with certain immunoassays. It is imperative that a dialogue regarding the use of herbal medicines be set up between physicians and their patients and that any reaction, be it positive or negative, be recorded and maintained on file. Extensive record keeping regarding a patient's health history and prior use of herbal remedies can then help health caregivers make better decisions regarding any medical treatments that they may prescribe.

In spite of the health risks inherent in some herbal remedies, more and more Westerners will continue to flock to them, and even use them in conjunction with other conventional drugs. What is to be done about this trend? One approach would be to further modernize herbal medicines by intensifying research in this field. For example, state-of-the-art technologies can be incorporated in herbal extraction and purification procedures to identify the bioactive compounds involved in a herbal remedy. Questions related to active ingredients, mechanisms of action, toxicology, and drug interactions will need to be satisfactorily addressed. Standardization according to all known bioactive components would be critical to ensure consistent pharmacological and clinical results. Finally, increased regulation of herbal medicines would be of tremendous benefit to the general public. Manufacturers of herbal remedies would be required to list the ingredients and ensure that they are pure. Dosage, possible adverse effects, and potential drug interactions should be listed as well. Such information would prove valuable not only to Westerners who choose to use traditional herbal medicines, but for most of the human race, who lack accessibility to modern medicine.

While this sounds like a reasonable approach to take, it will be extremely difficult to implement. One significant challenge will be the inherent difficulty in determining what active ingredients are present in some herbal remedies. Although experimental approaches have advanced rapidly over the last decade, the active ingredients of many herbal remedies are in fact very large and complex biological macromolecules that are not so easy to characterize using modern analytical chemistry. The timely and expensive process of analyzing these herbal remedies will continue to be a bottleneck in the use of natural products as medicines for some time to come. In addition, properly designed clinical trials are paramount to determine both how effective an herbal medicine is and what potential drawbacks it may have with respect to adverse reactions, toxicity, and interactions with other drugs. Pharmaceutical companies themselves pay dearly for properly managed clinical trials as well as for the opportunity to have their medicines examined and approved by the appropriate government regulatory body. As clinical trials are extremely costly ventures, it is unclear as to where the funding for the clinical studies of herbal medicines will come from. A few clinical trials are run by the government, as the NIH-based NCCAM can attest, but this avenue is quite limited with respect to funding and can only examine a small fraction of herbal medicines at a time.

What is the future of the battle between traditional herbal medicine and modernized Western medicine? In 2007, according to the NCCAM, 38.3% of adults in the United States used some form of complementary and alternative medicine. Indeed, efforts are being made in certain instances to combine traditional herbal medicine for which there is evidence of safety and effectiveness with conventional treatments. This integrative medicine approach may turn out to be much more cost effective for countries with burgeoning populations, such as China. Information that is collected regarding the parameters by which herbal medicines can be safely and effectively utilized can only be helpful to the 80% of the world's population that stands little chance of ever having access to Westernized modern pharmaceuticals.

CHAPTER 4

Farming Medicines from Plants

From the dawn of civilization, plants have been used as a food resource as well as a means by which to treat various illnesses. Specific plants, or plant tissues, have impeded infections, reduced inflammation, and even prevented the progression of certain cancers. Plants still remain a significant resource for modern drug discovery and are being actively used in alternative and complementary medicine today. Up until only the last few decades, bioactive compounds from which the majority of modern medicines originate were still being extracted and purified directly from plants. We now turn to the use of plants as production platforms and delivery systems for vaccines and other pharmaceutical proteins, a field that is rapidly gaining momentum.

Production of plant-derived biopharmaceuticals, or molecular farming as it is sometimes called, involves the use of biotechnology to produce pharmaceutical compounds such as vaccine proteins in plant tissue. What makes this chapter stand out from the earlier ones is that these therapeutic agents do not originate naturally in plants; rather, they have no plant origins whatsoever. Thus, instead of identifying novel plant compounds that can be incorporated into medical practice, molecular farming utilizes plants as production platforms to generate modern biopharmaceutical proteins. This spectrum of medicinal proteins produced in plants has grown substantially over the past few years and ranges in scope from the generation of human monoclonal antibodies against HIV, to vaccine proteins against anthrax and other biological warfare threats, and even to an assortment of anticancer therapeutic agents for the newly emerging field of personalized medicine.

The idea of molecular farming for vaccine proteins was first conceived in direct response to an urgent need from developing countries for effective and inexpensive therapeutic proteins. It was necessary that vaccines such as these be easily transportable to remote rural areas yet remain stable for great lengths of time in the absence of refrigeration. Plant-derived vaccines can be directly consumed in the form of edible plant tissue such as tomatoes, corn, or bananas and can effectively elicit an immune response to a particular pathogen. Furthermore, while many conventional vaccines currently in use must be kept under refrigeration and have limited shelf lives, vaccine proteins produced in plants are generally quite stable and can be stored as seed at room temperature for months or even years without substantial losses. Since rural communities generally maintain their own crops, plants expressing vaccine proteins can be raised using local farming techniques, and do not require sophisticated instrumentation for processing.

The key driving force of farming for vaccine and therapeutic proteins in plants comes from the desire to combat and control the most devastating childhood diseases found in developing countries. For example, diarrhea due to diseases such as cholera, Norwalk virus, and rotavirus remain major causes of infant mortality in the Third World today. Over 20% of the world's infants still do not have adequate access to vaccines, and infectious diseases are responsible for several million preventable deaths a year, due to constraints such as production and distribution of vaccines (Figure 4.1). For example, an enterotoxin released by some strains of *E. coli* is responsible for the annual deaths of 3 million infants, principally in impoverished areas. If a plant-derived vaccine to this enterotoxin was made available and administered to mothers in these locations, their children would have the opportunity to become immunized in utero by either transplacental transfer of maternal antibodies or through consumption of breast milk.

Plant-derived vaccines may also become a plausible wave of defense against infectious diseases that fall short of public awareness and are poorly financed, such as hookworm or dengue fever. In fact, several different vaccine proteins can be produced simultaneously in the same plant, so a person could potentially be immunized for several diseases at once by consuming the edible tissues of a particular plant.

Achieving immunization programs that are global depends greatly on the cooperation between governments and international health organizations. Philanthropic organizations such as the Gates and Rockefeller Foundations have collaborated with governments to push this promising

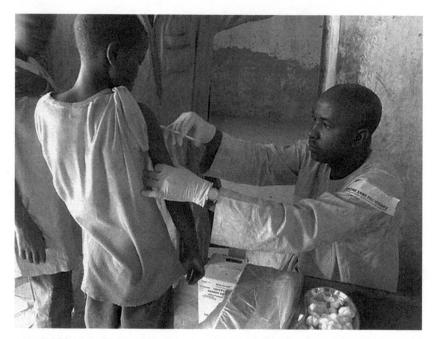

Figure 4.1.
Nigeria, meningitis vaccination campaign, April 2009, Katsina State, Musawa Local Government area. Vaccines are much needed in African countries. From early March 2009, Médecins Sans Frontières started vaccinating in several states in northern Nigeria. In one day as many as 50 teams were able to vaccinate in several locations. In 2009, meningitis cases were diagnosed earlier than usually in the season. MSF launched an emergency intervention to treat the affected patients and organize mass vaccination in nine states in northern Nigeria. This has proved to be the biggest vaccination campaign in Nigeria so far with a target of more than 5 million people to be vaccinated in this country only. Photo by François Servranckx /MSF.

new technology forward. In the early 1990s, the World Health Organization, in conjunction with a consortium of other philanthropic organizations launched the Children's Vaccine Initiative, an international endeavor to ensure that children throughout the world have access to immunization programs. Organizations such as these were instrumental in moving forward research with respect to plant-based vaccine development.

The first proof-of-concept studies will be discussed later in this chapter and involve the use of transgenic potato tubers expressing a hepatitis B vaccine. Potatoes, being relatively easy to genetically engineer, were consumed raw for these preliminary clinical trials to ensure that no vaccine protein was degraded through cooking. Other crops which predominate in many developing countries and are generally eaten raw, such as banana, are eventual target plants of interest. Banana fruit expressing vaccine proteins to cholera, rotavirus, and Norwalk virus, for example, could be pureed to maintain a consistent level of vaccine protein and offered to children as an effective means to ward off diarrheal diseases at a small

Figure 4.2.
Bananas are a potential food source for vaccines. Photo by Anson Eaglin, US Department of Agriculture.

fraction of the cost of a conventional vaccine counterpart (Figure 4.2). This strategy of growing vaccines and other therapeutic proteins in plants that are routinely eaten and can be farmed easily could potentially have a rapid and beneficial effect for those who need it most. It has been predicted that plants grown in a mere 40-acre plot of land can produce enough hepatitis B vaccine to vaccinate all of China each year, and a 200-acre plot could vaccinate all of the babies of the world.

Antibodies raised against rabies virus in plants present another example of the potential of this technology to make a significant improvement for the lives of many in developing countries. Rabies is not considered a disease of high priority in highly industrialized countries due to its low incidence of contraction, but the number of deaths due to rabies virus averages 55,000 a year in Southeast Asia and Africa. Since the vaccine currently available is very costly, antibodies against rabies virus expressed in plants inexpensively and in large quantities could present an attractive alternative. Antibodies such as these could be grown locally and stably stored at room temperature, without the need for sophisticated manufacturing equipment for downstream processing.

Recently, the European Community funded the Pharma-Planta project (www.pharma-planta.org). Scientists involved in this project signed a "Statement of Intent For Humanitarian Use" to focus on health needs of the poor in developing countries. Similarly, the Bill and Melinda Gates Foundation set up "The Grand Challenges in Global Health" initiative, which comprises a committee of medical experts to identify needle-free delivery systems and heat-stable preparations of vaccines. Molecular farming for vaccines in plants could very well be the answer that these organizations are looking for.

NEW PRODUCTION SYSTEMS

Since the initiation and development of genetic engineering in the 1970s, biopharmaceutical proteins have principally been produced by bacterial or yeast fermentation systems, or by mammalian cell culture. Over the past few years, transgenic animals (animals that harbor and express genes from other species) have also entered the arena for therapeutic protein production (Table 4.1). Transgenic animals such as mice, rats, rabbits, pigs, sheep, and cows are also producing and expressing in their milk pharmaceutical proteins such as insulin, blood anti-clotting factors, monoclonal antibodies, and rotavirus vaccines. Rabbits, for example, are a popular production system as they have a rapid cycle of pregnancy and maturation, and no known prion or viral diseases transmissible to humans.

Although it seems likely that some people may have concerns for the development of transgenic animals, there are other, more pressing needs for this technology. For example, donor rejection is a chief stumbling block for advances in the field of xenotransplantation, that is, the transplantation of an organ from an animal such as the pig into a human. Organs from transgenic pigs that have had donor rejection proteins replaced with human protein counterparts would have a much better chance of being well tolerated in a transplant patient.

While the use of plants as expression platforms for vaccine proteins is currently still in its infancy, plants have already proven to be highly advantageous for a number of reasons. One great advantage that plants have over conventional microbial fermentation systems is that plants are more "animal-like" in their ability to fold and assemble foreign proteins. This is

Table 4.1. EXPRESSION PLATFORMS FOR PROTEIN PRODUCTION

Expression System	Bacteria	Yeast	Mammalian	Transgenic Animal Products	Plants
Cost	Moderate	Moderate	Expensive	Expensive	Inexpensive
Safety	Contamination issues	Contamination issues	Contamination issues, infectious agents	Safe	Safe
Speed of production	Fast	Fast	Fast	Slow	Fast
Scalability	Moderate	Moderate	Moderate	Moderate	Rapid
Protein modifications	No	Yes	Yes	Yes	Yes

because plant cells have an endomembrane system that enables proteins to be modified in a manner that resembles proteins produced in animal cells. For example, complex proteins that consist of several separate components, such as complete antibodies, have been produced in plants and demonstrated to be just as biologically functional as antibodies produced in mammalian cell culture. Bacterial cells, which lack this endomembrane system entirely, cannot perform this essential process and thus are not able to produce complex multimeric molecules such as antibodies. Vaccine proteins can also be targeted to a plant cell's endoplasmic reticulum, a subcellular organelle with an environment favorable for protein folding, thus increasing the amount of vaccine protein produced that is biologically active.

Furthermore, working with plant cells substantially reduces the health risks associated with mammalian production systems, such as contamination from infectious agents. Proteins produced from mammalian cell culture must undergo both extensive and expensive purification pathways to ensure that no contaminants are carried along to the final product. Since plant-made vaccines, on the other hand, can be administered by direct consumption of plant tissue, only partial processing is required, saving both time and money for the manufacturer.

HOW DOES IT WORK?

It should be no surprise that some vaccines delivered in plant tissue are capable of eliciting a more powerful immune response than conventional vaccines. What gives plant-derived vaccines the leading edge? The portal of entry for many pathogens is through mucosal surfaces, such as the gastrointestinal tract, or gut. Mucosal surfaces consist of epithelial cells that line the surfaces of the gut and overlie an assortment of lymphoid tissues containing many different cell types involved in the immune response. In humans, the gut-associated lymphoid tissue (GALT), specific to the intestinal tract, represents 70% of the body's entire immune system. Peyer's patches, named after the 17th-century Swiss anatomist Hans Conrad Peyer and found to reside in the lower small intestine, are a component of this immune surveillance system and are directly involved in the mucosal immune response. Foreign antigens are processed and presented to different cells of the immune system at these Peyer's patches.

In general, most antigens cannot survive for any prolonged period of time within the harsh environment of the gastrointestinal (GI) tract without becoming degraded. This fact poses a significant obstacle for the delivery and presentation of antigens to the intestinal immune system. Vaccines that are made in plants offer a select advantage since the plant tissue itself has a protective effect and prevents rapid degradation of the antigen while it passes through the gut.

Another hurdle to overcome is that many antigens do not become recognized as foreign, and as a result, they cannot serve adequately as immunogens and elicit an immune response. One way to mitigate this hurdle is to include a substance that stimulates the immune system's response, such as an adjuvant, along with the vaccine protein. Cholera toxin subunit B, or CT-B, has been shown to elicit a strong immune response. Five CT-B subunits form a stable ring structure that stimulates a robust immune response in the intestinal mucosa upon oral administration (Figure 4.3). As a result of this, CT-B not only can function as an immunogen; it can also act as an efficient transmucosal carrier molecule and delivery system for plant-derived subunit vaccines. Proteins or portions of proteins that are only weakly immunogenic can thus be fused to CT-B and result in an immune response much greater than the antigen would elicit on its own. The increase in immunogenicity of the antigen is due to CT-B's ability to present it in a more optimized context to the mucosal immune system.

Besides CT-B, other vaccine proteins have been shown to successfully act as carriers for other antigens. Norwalk Virus (NV) coat protein, for example, when expressed in plant cells, has a tendency to self-assemble into virus-like particles (VLPs). These particles, known as NVLPs, closely resemble the fully infectious version of Norwalk Virus and are able to elicit a strong immune response when delivered orally, but are noninfectious as they lack any viral genetic material. Since oral and nasal delivery of NVLPs efficiently produce antibodies at distal mucosal sites, NVLPs

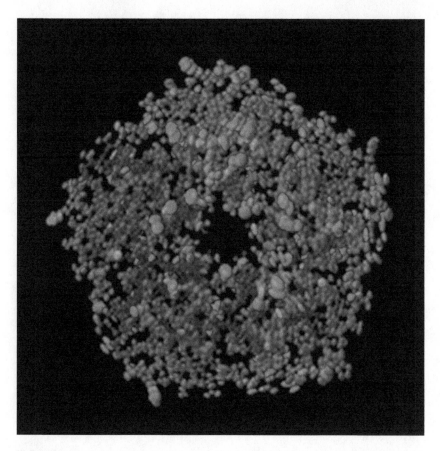

Figure 4.3.
Structure of the Cholera Toxin B-pentamer refined at 1.25 A. Copyright IUCRCrystallography
Journals Online (http://journals.iucr.org).

can also be used to deliver a specific antigen of choice in the form of a chimeric fusion protein and generate an immune response to that antigen.

TECHNOLOGIES USED TO DESIGN PLANTS EXPRESSING BIOPHARMACEUTICALS

There are two basic methods by which foreign proteins can be expressed in plants. One popular procedure is to stably transform plant cells and regenerate them into full-grown transgenic plants expressing the protein of interest. Another technique is to transiently express foreign proteins by infecting mature nontransformed plants, either using a technique known as Agroinfection or through a virus expression vector that harbors the gene encoding the protein of interest. Each methodology has its own

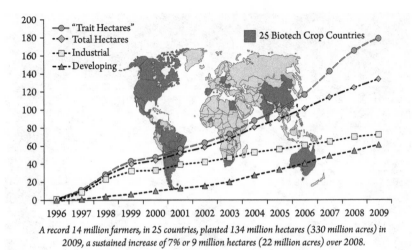

200
180
160
140
120
100
80
60
40
20
0

—○— "Trait Hectares"
--◇-- Total Hectares
--□-- Industrial
--▲-- Developing

■ 25 Biotech Crop Countries

1996 1997 1998 1999 2000 2001 2002 2003 2004 2005 2006 2007 2008 2009

A record 14 million farmers, in 25 countries, planted 134 million hectares (330 million acres) in 2009, a sustained increase of 7% or 9 million hectares (22 million acres) over 2008.

Figure 4.4.
Global area of genetically modified crops (million hectares, 1996–2009). From Clive James. (2009). *Global Status of Commercialized Biotech/GM Crops: 2009.* ISAAA Brief No. 41. ISAAA: Ithaca, New York.

specific advantages and disadvantages, and both can only be achieved through the use of modern plant biotechnology.

Transgenic Plants

There are few places on earth today where the term "transgenic plant" is unknown. Originally developed and introduced as crops with improved features such as pest resistance and herbicide tolerance, transgenic plants now exist with a wide array of attributes, from improved abilities to deal with environmental extremes, such as drought, high temperatures, and flooding, to possessing the capability to remove heavy metals and other pollutants from watersheds or soil, and even to exhibiting improved nutritional qualities. Table 4.2 provides a sample list illustrating the wide range of characteristics produced in these "first- and second-generation" transgenic plants. There is no doubt that the development of transgenic plants is one of the key hallmarks of modern biotechnology today. Initial genetically engineered crops such as soybean and corn first appeared in the United States in the mid-1990s. Since then, transgenic crops have been commercialized in many countries and over much of the world (Figure 4.4). Transgenic plants present enormous potential in regard to their use as bioreactors—in other words, as cost-effective and safe systems for the large-scale production of proteins for industrial, medicinal, and veterinary uses. Table 4.3 lists a small sample of biopharmaceutical proteins produced using transgenic plants as bioreactors.

Table 4.2. TRAITS OF FIRST- AND SECOND-GENERATION TRANSGENIC PLANTS

First-Generation Transgenic Plants	Second-Generation Transgenic Plants
Virus resistance	Biofortified plants with improved nutritional qualities
Herbicide tolerance	Drought-resistant, salt-tolerant, heat-tolerant plants
Insect resistance	Plants that phytoremediate polluted soils
Bacterial resistance	Vaccines and therapeutic proteins produced in plants

Table 4.3. EXAMPLES OF VACCINE AND THERAPEUTIC PROTEINS CURRENTLY MADE IN PLANTS

Disease/Condition	Antigen/Protein Produced	Plant Used
Enterotoxigenic E.coli (diarrhea)	LT-B	Potato, corn
Hepatitis B virus	HBsAg	Potato
Rabies virus	Spike protein	Spinach
Allergies	Japanese cedar pollen antigens	Rice
Non Hodgkin's lymphoma	Tumor antigens	Tobacco
Influenza virus	HA protein	Tobacco
Bubonic plague	F1-V	Tomato
Herpes simplex virus	Monoclonal antibody	Soybean

Transgenic plants are generated from regular plant tissue that has been transformed, that is, has acquired new genetic material. The most popular techniques by which plants undergo transformation events are *Agrobacterium*-mediated transformation or biolistic delivery. The former technique is based on the fact that the soil bacterium known as *Agrobacterium tumefaciens* can induce tumors in some plants by transferring a portion of its own genetic material into an infected plant cell which then becomes incorporated into the plant genome. The genetic material responsible for inducing tumors was identified in the 1970s as tumor-inducing plasmids, or Ti-plasmids. This discovery revealed the potential for use of *Agrobacteria* and their Ti plasmids as a means by which to introduce foreign genes into plant cells. Cells that become successfully transformed can then be easily regenerated into full-grown, mature transgenic plants, as depicted in Figure 4.5. The plants produced can be tested for altered characteristics, such as pest resistance, or for biological activity of the foreign protein produced. For example, the popular anticancer drug Taxol can now be synthesized and purified from transgenic tomato plants.

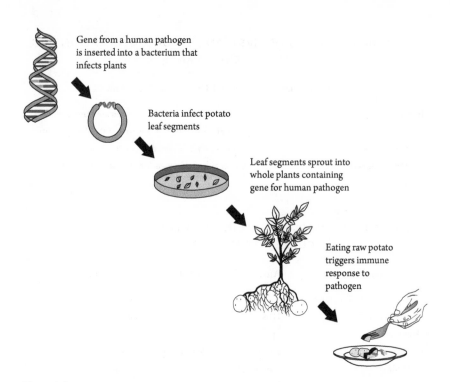

Figure 4.5.
(a) Steps involved in generating vaccines in plants. Figure courtesy of National Institute of Allergy and Infectious Diseases, National Institutes of Health.

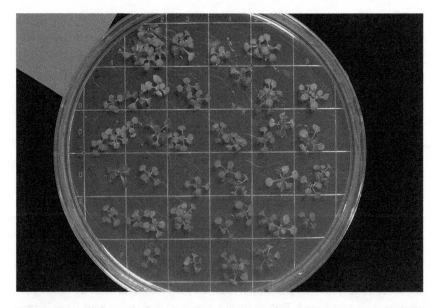

Figure 4.5.
(b) Plants growing on Petri dish. Photo courtesy of Linda Savage, Michigan State University.

Taxol, described in Chapter 2, originates from the Pacific yew tree; it is difficult and expensive to extract, and there are not enough trees available on earth to satisfy the requirements of all cancer patients who require it.

Biolistic delivery is another methodology developed to generate transgenic plants. This technique, which involves particle bombardment with a device known as a gene gun, was developed for plant transformation from the knowledge that many species of plants were not within the restricted host range of *Agrobacterium* and thus could not easily be developed into transgenic plants. For example, monocots such as rice or corn are not naturally infected by *Agrobacteria,* and as a result are often transformed by particle bombardment. In this procedure, high velocity microprojectiles (microcarriers consisting of subcellular-sized gold or tungsten particles coated with the desired DNA of interest) can be "shot" using a "gene gun" into plant tissue. The microcarriers penetrate the plant cells and the genes are released within the cells. Under optimal conditions, cell injury is minimal and the new genes can then be maintained within plant cells either as stable transformants or for transient expression studies.

Plastid Expression Systems

An alternative and very promising expression system for vaccine proteins stems from chloroplast, or plastid engineering. Chloroplasts are small organelles present in the hundreds or even thousands, depending on the type of plant cell, which maintain their own genetic material and remain distinct from the nuclear genome. It is the chloroplasts that contain the photosynthetic machinery of the plant cell—that is, the location where water and carbon dioxide are converted into sugar and oxygen in the presence of sunlight. Plant cells themselves have evolved from the original capture of photosynthesizing microbes known as cyanobacteria by a pre-eukaryotic cell. The two cells acted as symbiants, with the chloroplast providing and storing food in the form of sugars to be used for the rest of the cell's metabolism. Chloroplasts belong to a group called plastids, which includes the green chloroplasts (present in photosynthetically active tissues such as leaves), the red or yellow chromoplasts (present in fruit and flowers), and several other plastid types.

Plastids possess a number of unique features that make them attractive systems for vaccine production (Figure 4.6). The number of chloroplasts found in a single plant cell can be great, and each chloroplast contains several copies of the chloroplast genome. For example, the number of copies of the chloroplast genome in a single pea leaf cell can be up to 10,000; and up to 50,000 identical copies of plastid DNA can be

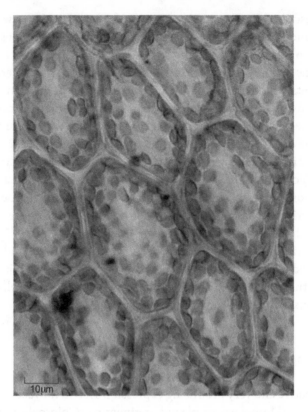

Figure 4.6.
Chloroplasts in plant cell.

found in a single wheat cell. Hence, the number of copies of a foreign gene of interest can be significantly greater (by the thousands) than the single or small number of copies that can be introduced into a nuclear genome. As a result, the amount of foreign protein produced in a plastid transformed transgenic plant can be quite enormous when compared to a standard, nuclear-transformed transgenic plant. Since chloroplast genomes resemble bacterial rather than eukaryotic genomes, they are easier to manipulate in many respects. A substantial advantage is that pollen contains no chloroplasts; therefore, there is no concern of plastid-transformed plants outcrossing with weedy wild relatives, a significant biological containment issue when dealing with transgenic crops in the field. However, to date, the number of species of plants that have success-fully undergone plastid transformation is more limited than those avail-able for nuclear transformation; also, since chloroplasts more resemble bacteria, they lack the protein-folding ability that nuclear-transformed plants possess. A number of vaccine and therapeutic proteins are now produced in transgenic plants via chloroplast transformation, including cholera toxin, interferon, anthrax protective antigen, plague vaccine, human somatotropin, tetanus toxin, and human serum albumin.

Transient Expression Using Virus Expression Vectors and Agroinfiltration

Although foreign protein expression using transgenic plants is fairly routine, several disadvantages exist with this technique, including the length of time required to grow transgenic plants and problems relating to low expression levels. Regeneration of a fully mature transgenic plant from a single transformed cell can take many months, depending on the plant type used. Occasionally, the design of the plant expression system can be confounded by the fact that the foreign protein produced actually exerts a toxic effect on the plant, resulting in fewer successful transformants being generated or plants consequently characterized by stunted growth. One approach that will avoid these problems altogether is the employment of plant virus expression vectors. Plant viruses can infect a host plant rapidly and express the desired foreign protein in a manner of days after inoculation, often at much greater levels than had the identical protein been expressed in a transgenic plant. Proteins that are toxic to the plant can also be more easily collected using this technique.

Two major groups of expression systems based on plant viruses have been developed for the production of immunogenic peptides and proteins in plants: epitope presentation systems (short pieces of protein, or peptides, which are fused to the plant virus coat protein [CP] and then displayed on the surface of assembled viral particles) and polypeptide expression systems (the entire unfused recombinant protein is expressed and accumulates within the plant separately from the virus particle). An example of a plant virus used as an epitope expression system is shown in Figure 4.7.

The use of plant viruses as expression vectors for biopharmaceutical proteins also comes with its own specific set of problems. For example, some recombinant plant viruses that harbor a foreign gene of interest tend to be less stable and can actually lose the foreign sequence by deletion through several passages in plants, making it difficult to conduct large-scale field trials. Other concerns regarding the application of plant viral-based vectors for molecular farming can include the intrinsically narrow host range of many plant viruses, as well as the plants' difficulties in tolerating adverse environmental conditions such as drought or extreme temperatures, and combating natural host defense responses. Some plant species have developed successful strategies to combat infection by plant viruses and other pathogens. Furthermore, although the

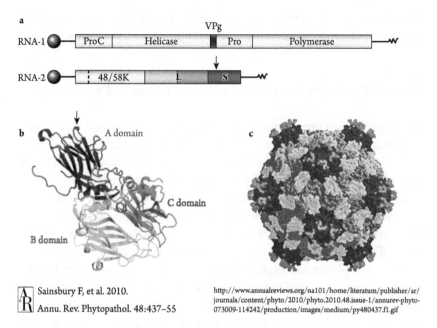

a

RNA-1 ⬤ ─[ProC │ Helicase │▮│ Pro │ Polymerase]─ ⋀

VPg

RNA-2 ⬤ ─[┊ 48/58K │ L │ S]─ ⋀

b

↓

A domain

C domain

B domain

c

Sainsbury F, et al. 2010.

Annu. Rev. Phytopathol. 48:437–55

http://www.annualreviews.org/na101/home/literatum/publisher/ar/
journals/content/phyto/2010/phyto.2010.48.issue-1/annurev-phyto-
073009-114242/production/images/medium/py480437.f1.gif

Figure 4.7.
Cowpea mosaic virus (CPMV) as a peptide-presentation system. The βB-βC loop of the small (S) coat protein (indicated by *red arrows* in *a* and *b*) is the most commonly used site for the insertion of foreign peptides. (*a*) Genome organization of CPMV RNAs. (*b*) Ribbon diagram of the icosahedral asymmetric unit, consisting of the two domains of the large (L) coat protein (*cyan and green*) and the S coat protein (*dark blue*). (*c*) Space-filling drawing of the CPMV capsid displaying an epitope from HRV-14 inserted into the βB-βC loop of the S protein shown in red. ProC, proteinase cofactor; VPg, genome-linked protein; Pro, 24K proteinase; 48/58K, movement protein; L, large coat protein; S, small coat protein. Panel *c* was kindly provided by Dr. T. Lin and Prof. J. E. Johnson, Scripps Research Institute, La Jolla, California, USA.

inability of plant viruses to harm people rules out the risks of human infection from exposure to the virus in the field or in food products, environmental concerns regarding biological containment of the virus itself remain. Recombinant viruses can possibly spread to weeds or nearby crops and their biopharmaceutical products could then potentially reside on both target and nontarget organisms. This problem can be dealt with by crop rotation, and genes that confer virus spread by insects or other vectors can be removed from the recombinant virus. Another simple solution is to ensure that all foreign protein production takes place under greenhouse conditions, where the virus is unlikely to escape.

Recently the biotechnology company Icongenetics, Inc. developed a technique using agroinfiltration for introducing plants with "modules" of recombinant virus vectors. Using this approach, a suspension of *Agrobacteria* harboring various virus vector components can be infiltrated using a vacuum into the intracellular space of all mature leaves of a plant. After this, the plants are then returned to the greenhouse where they grow and are later harvested for the biopharmaceutical protein. The virus expression vectors used in this case lack the gene encoding the coat protein and therefore cannot assemble into infectious virus particles, thus ensuring their inability to be transmitted to weeds and other nontarget organisms.

Agroinfiltration is the technique of choice used by many biotechnology companies today for biopharmaceutical production in plants. For example, Bayer Innovation GmbH, a subsidiary of Bayer AG, uses an agroinfiltration and plant virus vector-based expression system known as "Magnicon" to express therapeutic proteins from tobacco plants.

CLINICAL TRIALS OF THERAPEUTIC PROTEINS PRODUCED IN PLANTS

The mucosal immune response to vaccines can be vastly improved by delivering vaccine proteins directly to the intestinal mucosa through direct oral ingestion rather than with needles. The number of clinical trials using plant-made vaccines is rapidly growing. Many of the original trials were conducted using plant-based vaccines against infectious diseases that predominate in developing countries, such as hepatitis B virus and the infectious agents responsible for childhood diarrhea. When the outcomes of these preliminary trials were proven to be successful, there was motivation for the development of other plant-derived vaccines. One big incentive has been the pressing need for inexpensive and rapid solutions to the looming threat of biological warfare, in which smallpox, anthrax and the plague might be used, or as a means to stave off global pandemics, such as could be caused by the H1N1 influenza virus. Plant-derived vaccines have even become a role player in the revolution of personalized medicine, in terms of experimental anticancer vaccines, such as non-Hodgkins lymphoma. Here are a few examples.

Hepatitis B Virus (HBV)

Hepatitis B virus remains a significant cause of mortality in developing countries and is responsible for the development of chronic liver disease, including hepatocellular carcinoma. Since needles and refrigerated vaccines can be hard to come by in many regions of the world where HBV vaccines are most needed, an inexpensive, plant-derived oral HBV vaccine would be of enormous benefit. This strong potential initiated one of the first clinical trials involving plant-derived vaccines. In this preliminary trial, five grams of either peeled transgenic or nontransgenic potato tubers expressing hepatitis B virus surface antigen (HBsAg) were fed to mice once a week for 3 weeks. The mice that were fed a transgenic version of the tubers produced anti-HBsAg antibodies, whereas control mice fed a nontransgenic version of the potato did not. Mice that were fed yeast-derived HBsAg, the conventional vaccine, also were unable to produce specific antibodies to hepatitis B virus. There are several possible reasons that the potato-derived vaccine fared better at producing an immune response. First, the HBV vaccine antigen produced in plant tissues was able to assemble into virus-like particles (VLPs), which are far more im-munogenic than the single antigen produced in yeast. Second, the slow digestion of the potato within the gut may have prolonged the release of antigens, and this delay resulted in more antigens reaching the Peyer's patches, leading to the observed enhanced immune response.

Based on this initial success using an animal model, a second study involving a double-blind, placebo-controlled Phase 1 human clinical trial has also been performed. In this study, uncooked transgenic potato tubers were fed in 100 gram doses to previously vaccinated individual volunteers. Ten out of 16 volunteers who ingested three consecutive doses of the transgenic potato tubers exhibited increased levels of serum anti-HBsAg titres. This proof-of-concept study paved the way for the potential of plants to provide useful and much needed vaccines and other therapeutic agents.

Plant-made Vaccines to Treat Diarrheal Diseases

In the Third World, two of the most devastating diarrheal diseases are caused by Enterotoxigenic *E. coli* (ETEC) and Norwalk Virus (NV). ETEC alone causes 3 million infant deaths a year. To determine whether immune responses to these diseases could be generated using plant-derived vaccines, a clinical trial was designed that involved feeding raw transgenic potato or corn expressing either heat-labile toxin-B (LT-B), the antigen of ETEC or NV coat protein in three doses to adult volunteers. The vast majority of these volunteers were able to develop more than adequate immune responses to ward off these diseases, again implying the potential of plants to satisfy the urgent need for vaccines in the Third World.

Rabies Virus

Although rabies virus is not a major cause of mortality in developed countries, it is responsible for 50,000–60,000 deaths a year in places like Southeast Asia and Africa. Prevention against rabies includes both immunization by vaccination and, if one contracts the disease, by applying rabies-specific antibodies around the bite wound to neutralize the virus.

Unfortunately, rabies-specific antibodies are expensive, difficult to produce in large quantities, and generally in short supply on a worldwide basis. The production of inexpensive and safe monoclonal antibodies (Mabs) in plants would therefore be a significant benefit to global health.

In the early 2000s, anti-rabies human monoclonal antibodies were developed in tobacco plants and exhibited an anti-rabies virus neutralizing activity comparable to their commercial counterparts produced in mammalian cell culture. Since antibodies are complex, multimeric molecules, each component was independently expressed in a single plant. These plant-derived antibodies contain post-translational modification motifs that differed from similar antibodies produced in mammalian cells. However, the plant-derived Mabs were still able to fold correctly and the small changes had no negative effect on the neutralizing and protective efficacy of this Mab.

Influenza Virus

Influenza A virus, the causative agent of the flu, produces 300,000–500,000 deaths and 3 million–5 million hospitalizations annually. New strains of influenza A arise each flu season as a result of point mutations within the surface glycoproteins hemmaglutinin (HA) and neuraminidase (NA). These point mutations provide a means by which any new emerging virus strains can evade the host's immune system. Until recently, all influenza virus vaccines have been produced in chicken eggs, a process that is expensive and has a lengthy production time.

Over the past few years, tobacco plants have been developed that express a highly immunogenic form of the HA protein, comparable in strength to the commercially available egg-produced vaccine. Moving forward on the momentum of these initial successes, the Canadian biotechnology company Medicago began the large-scale development of pandemic and seasonal influenza vaccines in tobacco plants. Medicago's production system involves inserting the influenza virus genes of interest into plants using agroinfection. The process of identification of the antigen to completion of a vaccine takes only 30 days, instead of the routine 9 months to perform the same task using egg-based techniques.

Anthrax

Anthrax, caused by the bacterium *Bacillus anthracis,* is classified as a category A biological warfare agent. This acute and fatal disease is acquired by inhalation or ingestion of bacterial spores. Once considered a problem more or less limited to those who work with livestock, where it was readily transmitted through meat, wool, and animal skins, anthrax has most recently gained notoriety in the 9/11 bioterrorism attacks, where it was thought to be delivered through spores in the mail. Protective Antigen (PA), one of the proteins expressed by *B. anthracis*, is named for its preferred use as a vaccine and has been produced at high levels in tobacco chloroplasts. This chloroplast-derived PA protected mice challenged with a lethal dose of anthrax. The development of a plant-based anthrax vaccine demonstrated the great potential of this technology to quickly and effectively produce vaccines to protect the general populace against biological warfare agents.

Personalized Vaccines to Combat Cancer; Patient-specific Tumor Vaccines

Non-Hodgkin's lymphoma (NHL), the fifth most common cause of death in the United States, refers to a group of tumor diseases in which B-cells (antibody producing cells) of the immune system multiply uncontrollably and accumulate in the lymph nodes, bone marrow, and other tissues. Approximately 18,000 Americans between the ages of 60 and 65 are annually diagnosed with NHL. Until now, physicians have generally treated the disease with toxic chemotherapy.

The chief problem for managing this disease stems from the fact that the diseased B cells are not recognized as foreign by the patient's immune system. These degenerate B-cells of the NHL patient express a unique "idiotype," or distinctive specificity for a particular antigen that is suitable to be used as the perfect tumor marker. To address this, a new frontier in medicine, known as personalized immune therapy, has been developed. This form of personalized medicine involves providing the patient with tumor-specific vaccines that flag the tumor cells by interacting with their unique surface proteins. This enables the tumor cells to be recognized by the patient's immune system, which can then destroy the diseased B-cells while leaving healthy B-cells intact.

A plant-based, patient-specific tumor vaccine against non-Hodgkin's lymphoma was developed by Bayer Innovation GmbH in 2009 (Figure 4.8). To establish the safety and immunogenicity of plant-produced vaccines in NHL patients, tumor-derived idiotypic regions taken from individual patients were expressed in plants. The vaccine proteins were produced using a plant viral expression vector and then transferred to tobacco leaf cells using the agroinfection technique described earlier. These patient-specific lymphoma vaccines were then purified and inoculated back into the same patients. In a Phase I trial, the plant-based experimental vaccine

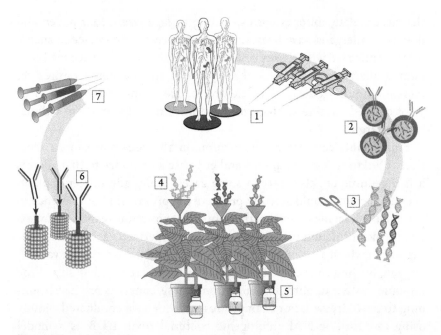

Figure 4.8.
Production of individual cancer vaccine. From Bayer Innovation GmbH (www.bayer-innovation.com).

triggered the immune systems to attack the tumors of 11 out of 16 patients with follicular B-cell lymphoma without any apparent adverse effects.

The beauty of this technique is the rapid speed and low cost by which the plant viral expression system can be used to express the required amounts of lymphoma vaccine for each patient. In this study, patient-specific vaccines were typically synthesized within 12 to16 weeks of receiving the biopsy specimens. Thus, vaccines produced by transient expression in plants could reach a larger population of NHL patients and quickly induce tumor-specific immune responses, making a significant impact on the population of patients assisted by these vaccines. If future trials prove to be successful, the experimental vaccine, synthesized quickly and cheaply, could become a short-term therapy administered immediately after diagnosis to keep tumors in check.

ALLERGIES, ORAL TOLERANCE, AND DOSE RESPONSE RELATIONSHIPS TO PLANT-MADE VACCINES

Thirty to forty percent of the population suffers from allergies in one form or another. Plant-derived vaccines have also been developed which induce oral tolerance to common allergies. For example, transgenic rice plants that accumulate epitopes corresponding to Japanese Cedar pollen and dust mite allergens have been shown to induce oral tolerance in such a way that mice fed with these plants became asymptomatic. Rice seed containing allergens can be stored for years at ambient temperatures yet remain immunogenic and efficiently induce immune tolerance. This plant-derived vaccine strategy has also been used to suppress asthma-based allergies.

Since plants can also produce common allergens that can be effectively used to reduce symptoms and generate a tolerance to that allergy, it is a genuine concern that in some cases, orally administered plant-derived vaccines could actually promote the progression of tolerance to a particular vaccine instead of the appropriate immune response. For example, the accidental consumption of a plant-derived vaccine may alter the way in which a given person would later respond to a similar vaccine antigen, leading to vaccine inefficiency as well as reduced ability of the immune system to eliminate infection. These concerns are just beginning to be addressed. For example, researchers have conducted a study using the corn-derived enterogenic bacterial toxin LT-B as a model system to find out the maximum nonstimulatory dose of an orally administered plant-derived vaccine antigen in mice. The scientists were able to establish a threshold level of orally administered plant-derived LT-B which did not stimulate detectable levels of antibody but could nonetheless induce immune priming, or a memory of that antigen in the immune system. This interesting relationship between oral administration of plant-derived antigens and its impact on the immune response is currently under hot pursuit and of great interest to plant biologists and immunologists alike.

THE SCALE-UP AND COMMERCIALIZATION OPPORTUNITIES FOR PLANT-DERIVED THERAPEUTIC PROTEINS

The rapid scalability and reduced cost for producing plant-made biopharmaceuticals have ignited significant attention in the corporate sector. While molecular farming may never completely take the place of many of the conventional vaccines that have been sold for years in industrialized countries, momentum is gathering in many sections of the world where specific needs can be addressed and adequately met, ranging from the delivery of affordable medicines for the poor to the prevention of global pandemics, and even to the rapid generation of personalized medicines.

Based upon the building evidence presented in this chapter, there are several reasons that the use of plants to produce vaccine and therapeutic proteins is becoming a particularly attractive concept. One significant advantage of the use of plant expression systems for molecular farming is the remarkable manner in which biopharmaceuticals can be safely stored in the form of corn kernels, wheat or rice grains, or in other seeds, without the need for refrigeration, and remain stable for months or even years in some cases, without substantial loss and with protein stability maintained. Crops that can express and be harvested for biopharmaceuticals are also more amenable to upscaling or downscaling by simply adjusting the acreage available. The huge quantity of generated plant biomass can compensate for any low protein yield that may be encountered with any specific plant production system.

Plant material can easily be processed on a large scale and many simple and inexpensive procedures to extract protein from plant tissues have already been developed. The extraction and purification of proteins from organisms or biological tissue can be a laborious and expensive process, and often represents the principal reason that vaccines and other therapeutic agents reach costs that become unattainable for many. Production of recombinant proteins in plants can cost as little as 2%–10% that of microbial fermentation systems and 0.1% that of the cost of mammalian cell culture systems. Since many of these plant products can be administered as food products, such as tomato juice expressing a vaccine against hepatitis B virus, for example, the time and expense of a purification process can be much reduced or virtually eliminated. This lower cost and improved speed of production should elicit interest from nations in developing countries, that would then have a feasible means by which to protect their own populations from preventable infectious diseases. Similar appreciation should extend to highly industrialized countries that desire to build biopharmaceutical production centers which are flexible enough to react quickly to a potential pandemic outbreak, such as one caused by influenza, or even to the threat of a biological agent used in a terrorist attack, such as anthrax.

The potential speed of production is another major advantage of vaccine production in plants. In many plant-based expression systems used to date, a product can be available for commercial use within a matter of a few months. For example, Biolex Therapeutics claims to have the capability to produce recombinant human therapeutic proteins in a small green aquatic plant called *Lemna* within several months. The rapidity of design, development, and production of new therapeutics using plant-based systems is particularly attractive for those interested in "personalized medicine," the production of small quantities of biopharmaceuticals such as the patient-specific monoclonal antibodies for non-Hodgkin's lymphoma described earlier in this chapter.

There are many types of crops and plant expression systems available for molecular farming. Plants can be grown as crops in the field or greenhouse, or even maintained in a cell culture in the form of a bioreactor. Each approach has its own advantages and disadvantages. One of the greatest advantages for the use of plant tissue culture over that of open field production is that most of the concerns such as transgenic plants outcrossing with their weedy relatives or issues regarding regulatory approval no longer have to be addressed. While variations in soil quality and weather patterns can make it difficult to implement good manufacturing practice conditions, which are indispensible for pharmaceutical production for field grown plants, cell suspensions can be grown in precisely controlled environments. In many instances, plant cell culture systems have been developed in which the biopharmaceutical protein product is secreted continuously from the plant cells, or by concentrating the therapeutic protein to the cell membrane and then purifying the membrane by simple fractionation techniques. Either example results in much reduced downstream processing costs. On the other hand, plant cell cultures also exhibit certain disadvantages such as the genetic instability of transformants when compared to field-grown plants. Other plant systems that have recently been developed for molecular farming include hairy roots, moss, and aquatic plant species, such as duckweed, kelp, and several species of algae (Table 4.4).

MOLECULAR FARMING REQUIRES ITS OWN UNIQUE SET OF REGULATORY GUIDELINES

Plants expressing vaccine and pharmaceutical proteins fall into the extraordinary position of being regarded as both medicine and potential food; thus they require a unique set of regulatory guidelines. In the United

Table 4.4. EXAMPLES OF PLANT SYSTEMS USED FOR MOLECULAR FARMING

Plant System	Examples
Crop Plants	Tobacco, potato, lettuce, carrot, rice, cotton, alfalfa, soybean, oilseed rape, tomato, corn
Cell Suspensions	Tobacco, rice
Hairy Roots	Tobacco
Physcornitrella Patens	Moss
Aquatic Plants	Kelp, algae, duckweed

States, molecular farming is under the regulatory authority of the FDA (Food and Drug Administration), the USDA (United States Department of Agriculture) and the EPA (Environmental Protection Agency), depending on the nature of the product and its intended use. As a result, the regulatory approval process of plant-made biopharmaceuticals is on average more extensive than the process by which traditional pharmaceuticals are currently overseen.

Products of molecular farming are examined and compared, whenever possible, to their bacterial or animal-derived counterparts. Issues such as the presence of potentially toxic or allergenic substances must be addressed to ensure that the product is safe and will not cause adverse affects in the patient. Any biological product, whether produced in plant, bacterial, or animal systems, must be fully examined for contaminating material. Plant material may possibly be exposed to various contaminants that would affect the purity and quality of the product, including agricultural chemicals, insects and other pests, dirt, pollen from other plants, and even fungi and bacteria.

CONCLUSIONS

Both low cost and ease of administration have made the concept of molecular farming a feasible solution for providing relief to Third World countries. The fact remains that infectious diseases are responsible for several million preventable deaths a year and approximately 20% of the world's infants are not immunized as a result of limitations on vaccine production, distribution, and delivery. Plant-derived vaccines could potentially play a lead role against those diseases that are less prominent and whose treatments are poorly financed, such as dengue fever, hookworm, and rabies. Moreover, the rapid response capability of a plant-based vaccine production platform makes this an attractive means by which governments can protect civilian populations from world pandemics such as one caused by H1N1 influenza. Along these lines, the US Department of Defense has shown great interest in developing a flexible technology that can quickly produce vaccines for threats such as biological warfare agents, and has funded the development of plant-derived vaccines to combat anthrax, for example.

Initially, researchers conjectured that plant-made pharmaceuticals could be produced in the field and directly eaten as a routine/local food source. Unfortunately, it soon became apparent that obtaining plant tissues containing the vaccine protein at high enough and consistent levels for oral consumption of a given population would not be easy. Scientists were forced to change their conceptions of a final plant-based vaccine product to one that will most likely be packaged as a capsule, juice, paste, or even perhaps as a suspension for oral delivery, rather than in the form of an entire fruit or vegetable.

The first USDA-approved plant-derived vaccine was granted to Dow AgriSciences in January of 2006; this was for a poultry vaccine against Newcastle Disease produced in tobacco cell culture. Many plant-derived vaccines and therapeutic proteins are completing the final phase of clinical trials and are due to be released within the next few years. For the world's poor, fast and affordable biopharmaceuticals produced by molecular farming couldn't arrive at a better time.

CHAPTER 5

Superfood

Functional and Biofortified Foods

Plants play an intricate role in nourishing us and helping us to ward off disease. The last three chapters have discussed in detail the medicinal function that plants can play in maintaining human health. This chapter describes the intimate role that plants have as a food source for improving human health and well-being in general.

It has become very clear that a diverse array of vegetables, fruits, and grains possess, in addition to vitamins and minerals, a large variety of biologically active compounds that act in a specific fashion to improve human health. The recognition of these bioactive components in functional food has recently gained much attention and publicity. As a result, functional foods offer additional benefits to human health which were previously unidentified.

The discovery and characterization of the health benefits of specific plants has also helped scientists to determine how to improve the nutritional properties of some crop plants which are consumed as staple foods in the diets of the world's malnourished. Biofortified foods offer a potential solution that could provide relief for many who reside in developing countries and have poor access to the essential vitamins and minerals that are critical to human health. Crop plants with improved abilities to uptake and accumulate minerals, or synthesize vitamins, amino acids, or oils are right around the corner.

The term "functional food," first introduced in Japan in the mid-1980s, refers to food which, in addition to being composed of basic essential nutrients, also contains ingredients that reduce the risk of chronic diseases, promote health, and extend longevity. Today, a great deal of evidence has accumulated to indicate that a plant-based diet can reduce the risk of chronic diseases such as cancer. An examination of results from 200 independent epidemiological studies clearly indicates that the risk of heart disease and cancer in people whose diets consisted of high amounts of fruits and vegetables was reduced to half that of people who consumed just a few of these types of foods. Studies have long shown that many phytochemicals protect plants, but only recently have we learned that they are also critical in protecting humans against diseases.

THE SCIENCE BEHIND FUNCTIONAL FOODS

The study of functional foods has become a rapidly developing area of food science research. Many plants used as foods naturally contain compounds beneficial to human health and have been demonstrated to play a direct role in preventing certain diseases. In fact, the identification of plants that contain numerous phytochemicals with potent anticancer and antioxidant activities has prompted the development of numerous new functional foods, many of which target specific health problems. For example, several epidemiological studies have suggested that the regular consumption of foods and beverages rich in flavonoids (a group of phytochemicals that will be discussed in more detail later) such as red wine, tomatoes, cocoa, and berries, is associated with a reduction in the risk of several conditions ranging from hypertension to coronary heart disease, stroke, and dementia.

Oxygen is required for sustaining the life of many higher organisms including humans, but it also plays a role in the damage of living tissue through the constant production of reactive oxygen species through a process known as oxidative stress. Oxidative stress is caused by an imbalance between the production of reactive oxygen and a biological system's ability to remove any reactive intermediates or easily repair the resulting damage. This results in damage to the cell, particularly to DNA. As a consequence, most organisms possess a series of antioxidant species that either prevent the formation of reactive species or remove them before the cell becomes damaged. Oxidation reactions can produce free radicals, which start chain reactions that damage cells. Antioxidants remove free radicals and prevent oxidation reactions from taking place (Figure 5.1). Antioxidants include reducing agents such as vitamin E, ascorbic acid, carotenoids, or flavonoids. Table 5.1 shows a list of bioactive compounds found in plant tissues.

Vitamin E refers to a class of chemical compounds found in vegetable oils that possess antioxidant properties. Among these, α-tocopherol is the best characterized. **Vitamin C** or **L-ascorbic acid** is an essential nutrient for a healthy diet and plays a role in a wide range of biochemical reactions in both animals and plants. Ascorbic acid is a highly effective antioxidant.

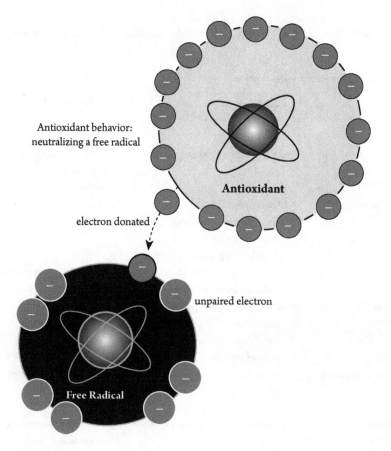

Antioxidant behavior:
neutralizing a free radical

Antioxidant

electron donated

unpaired electron

Free Radical

Figure 5.1.
Antioxidants neutralize free radicals. Original artwork ©2009 Joyce E. M. Wall, paisleypearl-press.comImage; used by permission.
J. E. M. Wall. (2009). Aging and antioxidants: finding a magic bullet in a biological battlefield of free radicals. Examiner.com.

Carotenoids are pigments naturally found in photosynthetic plants. Carotenoids play a role both in photosynthesis and as protection against potential cellular damage from sunlight. Many different carotenoids are found in food, and most have antioxidant activity. Not only are carotenoids effective as free-radical scavengers but they have also been shown to enhance the immune system. People with diets rich in carotenoids from natural foods, such as fruits and vegetables, are healthier and have lower mortality from a number of chronic illnesses than those who consume fewer of these elements. Among these, α-carotene, β-carotene, lycopene, and lutein are all carotenoids; β-carotene, a strongly colored red-orange pigment abundant in plants and fruits, is a precursor of vitamin A.

Flavonoids are a class of chemical compounds that are highly abundant and widely distributed throughout all plants. Flavonoids play many roles in plant

Table 5.1. LIST OF BIOACTIVE COMPOUNDS FOUND IN EDIBLE PLANTS

Bioactive Compound	Description	Function
flavonoid	Class of compounds widely found in plants, responsible for pigment	Antioxidant activity
phenol, polyphenol	Group of molecules widely found in plants	Antimicrobial activity, anti-proliferative effect on cancer cells
β-carotene	Class of compounds known as terpenoids, found in carrots, orange-colored vegetables and fruits, and green leafy vegetables such as kale	Primary source of vitamin A
lycopene	Bright red carotene found in tomatoes and other fruits and vegetables	Antioxidant, possible anticancer activity
lutein	A carotenoid found in green leafy vegetables	Antioxidant
phytoalexin	Class of chemicals found in a wide number of plants	Antimicrobial substances
terpene, terpenoid	Produced from a wide variety of plants	Chemical building blocks for a number of biological compounds
alkaloid	Found in a variety of plants and other organisms	A group of nitrogen-containing compounds, often have pharmaceutical properties for use as toxins, stimulants, anesthetics, etc.
phytosterol	Found in vegetable oils	Plant steroid alcohols, can lower cholesterol
phytic acid	Derived from a wide variety of plant tissues, a constituent of chlorophyll	Involved in vitamin E, K1 synthesis
squalene	Produced from a wide variety of plants	Part of the Mediterranean diet, possible anticancer activity
tocopherol	A variety of plant oils, including wheat germ oil and sunflower oil	Has vitamin E activity, may prevent Alzheimer's disease, heart disease, and cancer
monounsaturated fatty acid	Found in nuts, olive oil, other plant sources	Reduces low density lipoprotein cholesterol
resveratrol	Type of polyphenol found in the skin of red grapes, cocoa powder	Anticancer, anti-inflammatory effects, beneficial cardiovascular effects

TABLE 5.1. *(continued)*

Bioactive Compound	Description	Function
anthocyanin	Red, blue, or purple pigmented flavonoids, found in tomatoes, blueberries, cranberries, grapes	Has a sunscreen effect, powerful antioxidant activity, anti-cancer, anti-inflammatory effects, prevents neurologic diseases, diabetes
melatonin	Found in a wide variety of plants and other organisms	Affects mood, sleep, has antioxidant activity, anticancer effects
stearic acid, oleic acid, palmitic acid, lineolenic acid	Found in vegetable oils, seed oils	Fatty acid, has cardiovascular benefits
procyanidin	Found in many plants including apples, cinnamon, cocoa, grapes, cranberries	Class of flavonols which reduce risk of cardiovascular disease, high antioxidant activity

metabolism. For example, some of the most important pigments for flower coloration are flavonoids; they therefore have a role in attracting pollinators. Flavonoids also are involved in plant defense. Moreover, when consumed as part of plant tissue, such as in fruits, vegetables, coffees, teas, berries, nuts, cocoa, and grapes, flavonoids have been shown to have health benefits to humans in the form of anti-allergic, anti-inflammatory, antimicrobial, and anticancer activities.

Other phytochemicals found in plants are the **phytoalexins**. Phytoalexins include terpenoids, glycosteroids and alkaloids. Phytoalexins are phytochemicals produced by plants with a role in plant defense, such as response to stress conditions or invasion by pathogens including bacteria, insects, or fungi. In addition, these plant compounds can exhibit a number of properties beneficial to man, including antioxidant activity, anti-inflammation activity, cholesterol-lowering ability, and even anticancer activity.

FUNCTIONAL FOODS AND HUMAN GENETICS

While many of the bioactive compounds mentioned have been extensively studied for their ability to protect the body against oxidative stress by scavenging for free radicals, for example, the results of epidemiological studies and health claims tend to be less convincing. This high variation in human response to the use of functional foods to prevent chronic disease can be partially explained by polymorphisms present within the human genome itself. That is, genetic variation among different individuals can alter the efficacy of each nutrient with respect to its absorption, circulation, or metabolism. In fact, while it is entirely feasible that some individuals possess a genotype that enhances their ability to derive significant benefit from an increased intake of such foods, another segment of the population may in fact be at a disadvantage. Genes therefore influence how a body absorbs and uses nutrients. The unique genetic background of each individual person results in variations in the assortment of nutrients required for maintaining good health.

Differences in nutrient utilization can clearly be seen between isolated populations of people who have over time genetically adapted to cope with different local environments. An example of this can be found in the trait of lactose intolerance, the incapacity of some people to metabolize lactose, the sugar found in milk, due to their inability to produce the enzyme lactase in adulthood. Most mammals stop producing lactase after weaning but those humans who could drink milk had access to protein and other nutrients, helping them survive the severe climates of northern Europe and central Africa. As a consequence, more of their descendants survived and carried the alternative lactase gene. Many northern Europeans and central Africans retain the ability to drink milk, although roughly 70% of the world's population are lactose intolerant. Similarly, certain populations seem to possess a higher frequency of a genetic variant that leads to gluten intolerance and have adverse immune reactions to the presence of wheat products in their diet.

Genetic variation among populations can also be strongly influenced by the nutrient content of soils, and as a direct consequence, to crops that are grown and consumed. For example, the lack of availability of iron, selenium, or other micronutrients in the diets of populations in some regions of the planet may have consequentially resulted in the selection for genes that increase the uptake, storage, and use of these essential nutrients in peoples of these regions.

The influence of genetic variation on nutrition has become an exciting new field of study known as nutrigenomics. Nutrigenomics refers to the study of how nutrients interact with genes to alter metabolism. It has been found that certain dietary chemicals act on the human genome by altering gene expression. Some of these diet-regulated genes may play a role on the onset and progression of certain chronic diseases, including diabetes, cardiovascular disease, and cancer. Nutrigenomics has created much interest for its potential to prevent or combat chronic disease through changes in one's diet. Some anticipate that nutrigenomics will bring forward new disciplines such as personalized nutrition, however, it remains a science in its infancy.

THE MEDITERRANEAN DIET: THE IDEAL DIET?

What are the best foods to include in one's diet in order to maintain good health and prevent disease? Most of us have heard of the Mediterranean diet, but what is it and why has it drawn such attention? The most recent nutritional and epidemiological studies show that an ideal diet could in fact follow some aspects of the Mediterranean diet. People who descend from Mediterranean countries such as southern Italy, Greece, and Crete generally have a longer life expectancy and a lower risk of being diagnosed with particular chronic diseases, such as cardiovascular disease, metabolic disorders, and certain types of cancer. These characteristic health benefits which are associated with the Mediterranean diet are due to both rigorous exercise and a set of specific eating habits, including the significantly large intake of functional plant foods and beverages, such as fruits, vegetables, nuts, wine, and olive oil, altogether encompassing a great array of bioactive phytochemicals (Figure 5.2).

Vegetables are also an important source of phytosterols, the intake of which is associated with a reduction in serum cholesterol levels and cardiovascular risk. This could be of great importance in developed societies in which cardiovascular disease is a main cause of death. Fruits also provide fiber, as well as vitamins, minerals, flavonoids, and terpenes, many of which provide protection against oxidative processes. Indeed, vegetables, fruit, and olive oil are most likely responsible for the apparent protection offered by the Mediterranean diet against hypertension. Plant-based foods and olive oil may contribute to the health of the vascular system in general. Some components of the Mediterranean diet appear to help improve cognitive function and mood. In particular, some flavonoids (which are frequently found in vegetables and fruits) are believed to have antidepressant activity.

Health benefits arise not only from the Mediterranean diet itself; how the food is prepared can also play a direct role. For example, cooking

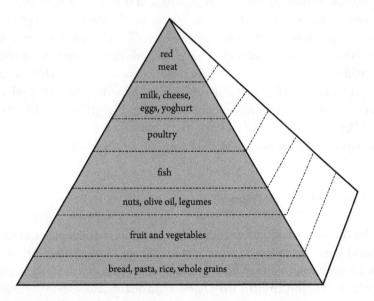

red
meat

milk, cheese,
eggs, yoghurt

poultry

fish

nuts, olive oil, legumes

fruit and vegetables

bread, pasta, rice, whole grains

Figure 5.2.
The Mediterranean diet, depicted as a food pyramid.

tomatoes in olive oil can greatly increase the absorption of lycopene, a carotenoid present in tomatoes that reduces the risk of certain cancers and heart disease. In this case, food preparation can influence the bioavailability of lycopene and other carotenoids. Furthermore, salads with various olive oil and vinegar dressings, and salads to which aromatic herbs have been added in the Mediterranean style have also been demonstrated to have a greatly increased antioxidant capacity. For example, lemon balm and marjoram have each been shown to significantly increase the antioxidant capacity of salad portions.

The Mediterranean diet highlights a number of functional foods, including olive oil, grapes, and grape products such as red wine, cocoa, berries, and tomatoes. The direct benefits of these functional foods is detailed in the following sections.

Olive Oil

Olive oil is the main source of fat in the Mediterranean diet. A great deal of evidence exists regarding its health benefits, particularly surrounding its ability to reduce the risk of heart disease and prevent several types of cancers. The benefits of olive oil extend to improvements in immune and inflammatory responses. Olive oil is known for its high levels of monounsaturated fatty acids and it is a good source of phytochemicals such as phenolic compounds. The content of minor components of an olive oil varies depending on the cultivar, climate, ripeness of the olives at harvesting, and the processing system used (for example, to produce extra virgin olive oil). Olive oil phenolic compounds have been well studied. In in vitro and ex vivo models, olive oil phenolics have been shown to possess significant antioxidant properties; they are also able to prevent endothelial dysfunction that can contribute to diseases such as hypertension. Also, olive oil phenolic compounds inhibit platelet-induced aggregation. In animal models, olive oil phenolics retained their antioxidant properties in vivo and delayed the progression of atherosclerosis (ischemic heart disease).

Mediterranean Herbs and Spices

Garlic, onions, herbs, and spices regularly used as condiments or in sauces or salad dressings may also behave as functional foods. Garlic and onions contain allicin, which may have cardiovascular benefits and help improve cognitive function. Allicin may have antioxidant and anti-inflammatory activities as well (see Chapter 3 on herbal medicines). Other herbs common to Mediterranean cuisine are also full of beneficial phytochemicals. Oregano is high in antioxidant activity, as is basil. Rosemary contains a number of potentially biologically active compounds, including antioxidants. Another common additive is the caper, which is found all over the Mediterranean basin and is consumed in salads or on pizzas (Figure 5.3). The caper plant itself is a perennial bush that bears rounded leaves and pinkish-white flowers. The plant is best known for the edible bud and fruit (caper berry), which are usually pickled, then eaten. Capers, which are full of flavonoids and other bioactive compounds, have been used in traditional medicine since ancient times and have recently been shown to have both anti-inflammatory and antioxidant effects.

Figure 5.3.
Capparis spinosa (caper bush). Painting from 1885, by Otto Wilhelm Thomé.
Source: Flora von Deutschland Österreich und der Schweiz.

Red Grapes

No better is the intertwined relationship between the chemical diversity of a particular food and the array of its biological activities symbolized than in the grape (Figure 5.4). Some of the beneficial effects of the Mediterranean diet with respect to human disease have been directly attributed to the polyphenols found in red wine. Different grape polyphenols work in a variety of ways to prevent cardiovascular and other inflammation-related diseases. The variety of bioactive compounds largely depends on where in the grape they are found. Grape seeds, grape skin, and grape juice contain several types of polyphenols, including resveratrol, phenolic acids, anthocyanins, and flavonoids. The antioxidant activity of grape polyphenols helps them to slow or prevent cell damage caused by oxidation. Grape

Figure 5.4.
Red grapes. Photo by US Department of Agriculture.

polyphenols can decrease the oxidation of low-density lipoprotein cholesterol, ("bad" cholesterol), thereby reducing one's risk of developing of atherosclerosis. Grape polyphenols can reduce blood clotting, abnormal heart
rhythms, and blood vessel narrowing. The exact mechanism by which the
wide range of health-promoting effects in grapes occurs remains to be elucidated. The antioxidant activity of these compounds may also be responsible for the cell-protecting ability of red wine reported in some studies.
Wine exercises its protective effect by inducing changes in the lipoprotein
profile, platelet aggregation, and endothelial function. Red grapes, red wine,
and even purple grape juice all have positive effects on endothelial function. Resveratrol has been shown to protect the heart and the kidneys.
Recently, importance has been associated with dietary indoleamines, mela-
tonin, and serotonin in red grapes. Much of the evidence regarding grape
polyphenols comes from laboratory experiments, although new studies
reinforce the disease-preventing benefits of grapes in humans. For example,
patients treated with grape seed extracts have shown improved cholesterol
levels, and other studies demonstrated that drinking Concord grape juice
can help to slow coronary artery disease as well as lower blood pressure in
patients with hypertension.

As described, consumption of red wine and other grape products may be
beneficial in a variety of ways in preventing the development of chronic degenerative diseases such as cardiovascular disease and cancer. It is likely that
regular and prolonged moderate drinking of red grape products positively
affects endothelial function. The beneficial effects of wine on cardiovascular
health would, of course, be greater if associated with a healthy diet. Moderate
red wine drinking, such as two eight ounce glasses for men and one glass for
women per day, may be beneficial to patients with cardiovascular disease.
Those health effects disappear very quickly, however, and serious health implications can occur when drinking is abused. Since studies indicate that
most of the beneficial effects of drinking red wine are attributable to the
polyphenols present in grapes, we may conclude that a diet that includes
grapes as well as other fruits and vegetables containing polyphenols is best.

Cocoa and Chocolate

Chocolate has long been considered to fall under the category of functional
foods. The unique fatty acid content of cocoa butter and the high prevalence of polyphenols make cocoa an ideal functional food. Both cocoa and
chocolate are extracted from the seeds, or cacao beans, of *Theobroma cacao*,
a tree native to South and Central America (Figure 5.5). The beans are harvested, fermented, dried, roasted, and ground into a "cocoa mass," which is

Figure 5.5.
Cocoa bean. Photo by US Department of Agriculture.

further processed to make a powder. Cacao bean products possess a unique chemical composition, half of which is cocoa butter, which in turn is composed of stearic, oleic, palmitic and linolenic acids. Approximately 15% of the dry weight of the cacao bean is composed of polyphenols.

The evidence that dark chocolate helps protect against heart disease and stroke is substantial. Dark chocolate contains phytochemicals such as flavonoids and works through various modes of action including prevention of blood clots, protection against oxidative stress, and lowering of blood pressure. Overall, chocolate should be considered a functional food for reducing the risk of cardiovascular disease.

Many studies illustrate the health-related benefits of cocoa. For example, studies in which flavanol-rich cocoa was given to overweight patients for 12 weeks resulted in a lowering of blood pressure and enhanced glucose metabolism. Additional studies have shown that such benefits are dose related and attainable with modest flavanol intakes from chocolate. Recent studies link cognitive decline to impaired cerebral vasodilatation and show that consuming flavanol-rich cocoa can increase cerebral blood flow. Cocoa has recently been shown to reduce platelet activation, which further suggests that it can play a role in reducing the risk of heart disease. Cocoa flavonoids can also display anti-inflammatory activity by modulating lymphocyte and macrophage activation. Studies using rats have demonstrated the use of cocoa to down-regulate the immune response, a property beneficial in controlling hypersensitivity and autoimmunity.

In one of the latest chocolate-related studies to be published, Spanish researchers found that cocoa powder enriched with procyanidin, a class of flavonol, significantly lowered blood pressure in rats suffering from hypertension. In fact, the cocoa product worked just as well for lowering the rats' blood pressures as a popular antihypertensive medication. Recently, a number of nutraceutical manufacturers have been boosting chocolate's flavonoid levels even higher, creating new chocolate products that are very rich in total procyanidins. A natural flavonoid-enriched cocoa powder, commercially known as CocoanOX, has been demonstrated to possess an antihypertensive effect in pre-clinical trials. A new soluble cocoa fiber product, obtained from cocoa husks and rich in soluble dietary fiber and antioxidant polyphenols, has been identified and its potential health effects studied in an animal model of dietary-induced hypercholesterolemia. This new cocoa product diminished the negative impact of the cholesterol-rich diet and resulted in lower food intake and body weight gain in comparison with control groups consuming cholesterol-free or cholesterol-rich diets with cellulose as dietary fiber.

Berries

In the search for functional foods, one of the most interesting revelations has been the high source of phytochemicals found in berries. Research has overwhelmingly established that berries have a favorable impact on human health. Berries are consumed by many cultures all over the world, are high in antioxidant activity, and display other health benefits as well. For example, cranberries have been used as both a food and a medicine for thousands of years (Figure 5.6). Cranberries are native to North America and Native Americans used them regularly to treat bladder and kidney diseases. Early settlers from England learned to include the cranberry in their diets, for its high vitamin C content and its use in combating digestive problems.

Cranberry is a well known treatment for urinary tract infections caused by *Escherichia coli* (*E. coli*). Originally, scientists thought cranberry worked by acidifying urine sufficiently to kill bacteria. Now, studies show that cranberries contain proanthocyanidins (responsible for cranberries' vibrant color), which prevent bacteria from adhering to the walls of the urinary tract, thus reducing the chance of infection. This is the proposed mechanism by which cranberries inhibit stomach ulcers caused by *Helicobacter pylori* and protect against gum disease and dental caries. A number of studies have suggested that cranberries and other dark-colored berries such as blueberries exhibit potentially beneficial effects in the treatment of diabetes, memory loss, neurodegenerative diseases of aging, radiation protection, and possibly heart disease, as well as reduce the risk of cancer.

Figure 5.6.
Cranberry harvesting. Photo by US Department of Agriculture.

Cranberries are rich in phytochemicals, such as phenolic acids and flavonoids. Cranberry fruit is high in antioxidants, partly from the presence of proanthocyanidins. A growing body of evidence suggests that cranberries and wild blueberries lower the risk of atherogenesis as well as cardiovascular disease by reducing blood pressure and inhibiting platelet aggregation; they are also known to have other anti-thrombotic and anti-inflammatory mechanisms. Furthermore, researchers have demonstrated that a blueberry-rich diet may improve stroke outcomes in rats and that blueberry and cranberry proanthocyanidins may assist in controlling tumor formation and the severity of some cancers.

Tomatoes

Tomatoes are a valuable source of nutrients, including β-carotene, vitamins C and E, and lycopene, a potent antioxidant that is responsible for the fruit's red color. As mentioned previously, incorporation of tomato products, particularly the tomato peel, in refined olive oil has been shown to be a very effective way to enhance lycopene, β-carotenoids, and other antioxidants in foods. Therefore, cooking tomatoes in olive oil—in a sauce, for example— enhances the bioavailability of carotenoids and phenolic compounds.

There is emerging experimental evidence supporting the correlation between an increase in tomato consumption and a reduced risk for both cardiovascular disease and prostate cancer. Recent animal studies examining tomato consumption, lycopene, and cancer suggest that diets containing tomatoes may indeed decrease the risk or the progression of prostate cancer. In a clinical study of more than 47,000 men, those who ate tomato-based products 10 or more times per week had less than one-half the risk of developing advanced prostate cancer than those who did not. The risk of a number of other cancers including breast, lung, and skin has also been inversely associated with lycopene levels. Proposed mechanisms by which lycopene could influence cancer risk are related to its antioxidant function.

The popularity of tomatoes in the diet of many industrialized nations and their role as a highly effective functional food have led to studies that aim to increase the level and bioavailability of the beneficial phytonutrients tomatoes have. Many of these are phenolic compounds that share common precursors with lignin, a component found in the cell walls of plants. To this end, one group of researchers was able to reduce lignin biosynthesis in transgenic tomato plants so as to enhance the availability of these precursors and, thereby, stimulate the production of soluble, potentially health-promoting, phenolic compounds in tomato. As anticipated, the level of these products was higher in the stems and leaves of the transformants as compared with the control (nontransgenic) plants. This was associated with an increased antioxidant capacity of the corresponding plant extracts.

Other researchers have used genetic engineering to increase carotenoid biosynthesis and accumulation in tomato plants; this has resulted in the generation of transgenic varieties of tomato containing high lycopene and β-carotene contents. Scientists soon discovered that this is not a simple process. An array of intrinsic regulatory mechanisms are clearly in place that remain in a "silent" state until manipulation of the pathway is initiated, and they offset attempts at increasing levels of phytochemicals in the plant. New engineering strategies, including systems and synthetic biology, are at present being employed to more efficiently enhance production of these bioactive compounds further.

DO DIETARY SUPPLEMENTS WORK?

After discussing the various plant compounds found in functional foods and the health benefits attributed to them, it would seem to be a foregone conclusion that taking such phytochemicals in the form of dietary supplements would be just as advantageous. However, this does not seem to be the case in many instances. β-carotene, for example, seems to lose its health-promoting effect when taken in supplement form alone. The reasons that individual phytonutrients contribute more effectively when in the context of other food components remains unclear at present and is the source of much speculation today.

So far, this chapter has described functional foods containing bioactive compounds that have health benefits or can reduce the risk of chronic diseases. While many functional foods may in fact be available as conventional foods that are part of a regular diet, some plants or plant products can be altered or processed in some way to increase their nutritional benefits. One example is the enhancement of antioxidants through metabolic engineering.

Anthocyanins are red, purple, or blue pigments that belong to the family of flavonoids and are produced by all higher plants. Blueberries, blackberries, and raspberries are all high in anthocyanin content. Anthocyanins have been shown to act as a "sunscreen," protecting plant cells from light damage and providing a powerful antioxidant activity. While dietary consumption of anthocyanins has been associated with protection against a broad range of human diseases, anthocyanin levels in the most commonly eaten fruits and vegetables are generally too low to confer optimal benefits. Recently, a research group generated a tomato plant that produced anthocyanin by expressing two transcription factors (which regulate gene expression) from the ornamental flower snapdragon in the plants. Tomato fruit produced from this plant were an intense deep purple in color and were found to have an enhanced antioxidant capacity, useful in fighting both cancer and chronic disease (Figure 5.7).

Figure 5.7.
Cancer-fighting purple tomatoes. Photo courtesy of John Innes Centre (jic.ac.uk).

Moreover, mice that were cancer susceptible and who were fed a diet supplemented with the high-anthocyanin tomatoes demonstrated a 30% extension of life span, further suggesting that these tomatoes have additional health-promoting effects.

Plant scientists have also successfully been able to modify fatty acid accumulation in seed storage oils. Although plant lipid metabolism in general is very complex, research groups are working on producing nutritionally enhanced vegetable oils with some degree of success. An example of one of these "designer oilseeds" is a transgenic plant expressing omega-3 fatty acids, the predominant fatty acid present in fish oils but not normally synthesized by higher plants.

There are other means by which to enhance functional food products without using genetic modification. For example, novel processing techniques can be used to improve conventional food products with superior health benefits. Products such as CherryPharm's natural health drink is one example of how a functional food such as tart cherries, with their high levels of anthocyanins and other beneficial phytochemicals, can be processed further into a highly potent functional superfood (Figure 5.8). Each eight-ounce bottle contains (in addition to apple juice and water) the juice of 50 tart cherries which have been processed in such a manner that the phytochemicals responsible for these antioxidant activities are retained. Polyphenolic antioxidants such as those found in health drinks like Cheribundi are considered premier disease fighters and protect the body against free radicals that ultimately cause cell damage leading to chronic and degenerative diseases. A number of clinical studies conducted using CherryPharm products have demonstrated beneficial effects for diverse inflammatory-related health conditions, ranging from rapid muscle recovery after strenuous exercise to minimizing the effects of arthritis, to lowering the risk of factors related to heart disease such as high cholesterol, and even to playing a role in improving the quality of sleep.

The production of new functional superfoods is just in its infancy. Table 5.2 provides a few more examples of such foods which will soon be on their way to the marketplace.

BIOFORTIFIED FOODS AND HIDDEN HUNGER

While food is more than abundant in affluent countries, such is not the case for many parts of the world. As these words are written, 1 billion, or 1 in 6 people on this planet, do not get enough food to eat. In addition, more than half of the world's population are suffering from a "hidden hunger," malnutrition due to the lack of essential micronutrients in their daily diets. For

Figure 5.8.
Cheribundi, a tart cherry juice developed at Cornell University by researchers in food science. Photo by Julie Prisloe (prisloephotography.com).

these people, meals tend to center around one or two staple crops, such as rice, maize, or cassava, and they do not have access to the variety of fruits and vegetables essential to a healthy diet. At present, deficiencies in nutrients such as iron, zinc, and vitamin A can account for close to two-thirds of childhood deaths worldwide. Malnutrition can also greatly impair an individual's physical and cognitive abilities. This results in a greatly reduced work capacity, which in turn imposes severe costs on a country's workforce. Moreover, according to United Nations (UN) projections, the world's population will increase from 6.1 billion in 2000 to 8 billion in 2025 and to 9.4 billion in 2050. Most (93%) of the growth will take place in developing countries. This rapid growth in population will present even greater challenges for governments regarding food security and the global problem of hidden hunger. Under current conditions, these governments

Table 5.2. EXAMPLES OF SUPERFOODS PRODUCED USING BIOTECHNOLOGY

Food	Improved Attribute	Method of Production	Mechanisms of Action	Health Benefit
Vegetable oil	Tocopherol (Vitamin E activity)	Transgenic plants	Expresses gamma tocopherol methyl-transferase	Reduces risk of heart disease, Parkinson's disease, Alzheimer's disease
Seeds	acetyl L-carnitine	Transgenic plants	Expresses carnitine acetyl-transferase	Reduces risk of Alzheimer's disease, depression, schizophrenia
Sunflower	erucic acid	Transgenic plants	lipid metabolism protein	Reduces risk of cardio-vascular disease, aids in skin care
Grasses	Essential amino acids	Transgenic plants	improves the nutritional content of plant seed	Improved animal diet without the need for additional supplements

can expect a higher increase in morbidity and mortality rates, lower worker productivity, and higher health care costs, all factors diminishing national economic development. In this way, the deleterious effect of malnutrition on human health and performance bring about a repeating cycle of poverty and economic dependency. Solutions must be set in place to prevent the increased duress that people from these regions will soon be facing. Since providing vitamin and mineral supplements to the rural poor in remote areas can be both expensive and difficult, the introduction of more self-sustainable, nutrient-rich, biofortified staple crops might be a step toward alleviating malnutrition in developing countries. The scientific evidence that has accumulated to date shows that technically it could be feasible for farmers in developing countries to grow biofortified crops with increased nutritional benefit in a self-sustainable manner.

It is important to realize that while minerals and vitamins are both components of micronutrients, they are substantially different. Minerals are inorganic compounds that cannot be synthesized from other molecules, and so plants secure them from the environment. To become more mineral-rich, biofortified plants must be persuaded to remove minerals from the soil and store them in their edible tissues. Vitamins, on the other hand, are organic molecules and can be synthesized from basic compounds found in life. For plants to have enhanced vitamin content, therefore, they must be provided with the capacity to synthesize these vitamins. Biofortification offers a cost-effective and sustainable strategy for dealing with micronutrient deficiencies, whether from a vitamin or a mineral source.

Biofortification can be accomplished either through conventional selective breeding techniques or through genetic engineering technology. In the former, varieties of plants are selected for the desired traits—for example, high iron content—and then cross-bred with each other. The progeny, which will contain both desirable and undesirable traits, are then "backcrossed" with other plants over several generations to select for the desired trait. New "hybrid" plants which then arise from these back crosses will contain this trait. In the case of genetic engineering, a new trait can be introduced via the insertion of a particular gene into the chromosome of a plant cell. Cells that have been successfully transformed can then be regenerated into mature transgenic plants.

While people require a nutritionally balanced diet with an adequate concentration of bioavailable micronutrients to prevent malnutrition and ward off disease, plants also require an appropriate level of micronutrients to resist disease and tolerate environmental stresses. Therefore, the efficiency in a plant's uptake of minerals from the soil is directly related to that plant's disease susceptibility. As a result, breeding plants for micronutrient uptake efficiency will also produce plants with improved resistance to diseases. For example, plants that are more micronutrient efficient tend to grow deeper root systems in mineral-deficient soils and are better at tapping subsoil water and minerals. Plant roots that are efficient in extracting surrounding external minerals are more disease resistant and can better penetrate deficient subsoils to make use of the moisture and minerals contained in subsoils. Increasing the micronutrient content of seeds increases seedling vigor and viability, thus improving their growth performance and yield when the seeds are planted in micronutrient-poor soils.

In the case of field crops, it is not enough that the plant itself contains appropriate concentrations of a particular vitamin or mineral. The micronutrient also has to be readily bioavailable, or taken up and used by the body for regular physiological functions as well. For example, the proportion of iron available in spinach leaves that is absorbed within the gastrointestinal tract and reaches the bloodstream would be a determinant of bioavailability. Some plants contain anti-nutrients or inhibitors, such as phytic acid, which reduce micronutrient bioavailability. This can be overcome by downregulating the concentration of an anti-nutrient in a particular biofortified food, and also by instructing that other foods known to contain such inhibitors are not included in the same meal with the biofortified food. Even the food preparation technique used (such as cooking, fermentation, or other processing techniques) can affect the level of bioavailability of a particular micronutrient.

A number of biofortified crops have been developed with the specific aim of relieving global malnutrition. A few examples are provided next.

Vitamin A(ß-carotene) Biofortification

Vitamin A deficiency is very much a global issue. Today, approximately 3 million preschool-age children have visible eye damage as a result of vitamin A deficiency. Annually, an estimated 250,000–500,000 preschool children go blind from this deficiency, and about two-thirds of these children will die shortly afterward. Besides its importance in vision, vitamin A has been shown to play an essential role in immune function. The precursor molecule β-carotene—a pigment found in many plant tissues, but unfortunately, not in cereal grains—is required for people to synthesize their own vitamin A. Administration of vitamin A capsules to infants and preschool children has been shown to help reduce mortality rates from all causes by 30%, and administration of capsules with vitamin A or β-carotene to women of childbearing age can reduce maternal mortality related to pregnancy by 40%.

Golden Rice is one very attractive biotechnological approach that offers a sustainable solution to reduce the prevalence of vitamin A deficiency-related diseases. To produce β-carotene in Golden Rice, two genes from other organisms were inserted into the rice genome to reconstitute the carotenoid biosynthetic pathway. This intervention leads in turn to the production and accumulation of β-carotene in the grains, the intensity of the golden color indicating the dense concentration of β-carotene.

Golden Rice represents the first nutrient-rich variety of crop that is sustainable as well as agronomically identical to locally adapted varieties and could become a powerful tool against malnutrition (Figure 5.9). Crops such as Golden Rice would have the potential to reach remote rural populations—those who cannot afford a diet rich in essential nutrients and who lack access to supplementation programs. The goal is to eventually provide the recommended daily allowance of vitamin A in what corresponds to the daily rice consumption of children in societies who depend largely on rice, such as India or Bangladesh. Golden Rice could complement children's diets in other countries as well, reducing clinical and subclinical vitamin A deficiency-related diseases. In fact, some researchers have shown that even modest additions of vitamin A can have substantial benefits to the diets of poor children who have serum retinol levels that are just below acceptable for normal growth and health. Small improvements such as these could be managed in a sustainable manner by the food system itself, and could provide beneficial consequences that are long lasting.

Figure 5.9.
Golden Rice. Photo courtesy of goldenrice.org.

Also in the pipeline is high β-carotene maize, an approach to alleviating the problem of vitamin A deficiency in Africa. Maize is a crop traditionally used in Africa and is often ground into a flour for traditional cooking and fermenting processes. The ability of β-carotene to be retained during traditional maize processing supports the feasibility of maize biofortified crops to alleviate vitamin A deficiency.

The Golden Rice project is often referred to as the model prototype for the potential of agricultural biotechnology to produce biofortified foods that can have enormous humanitarian benefit. Freedom to operate was maintained from the beginning and the technology was to be provided free of charge with no attached conditions to subsistence farmers in developing countries. In spite of the potential to eliminate what many would consider one of the largest causes of malnutrition in the world, Golden Rice continues to draw criticism from anti-genetic-modification (GM) groups, who worry that acceptance of GM technology involved in generating these plants would open the floodgates toward a general global acceptance of other genetically modified foods.

Iron Biofortification in the Developing World

Iron deficiency is the most common and widespread micronutrient deficiency in the world today. According to the World Health Organization, more than 2 billion people suffer from iron deficiency. Iron deficiency during childhood and adolescence impairs both physical growth and mental development. In adults, iron-deficiency anemia reduces the ability to perform physical labor effectively. Iron deficiency increases the risk of women dying during delivery or during the postpartum period. Based on the recorded incidence of anemia, most preschool children and pregnant women who reside in developing countries and at least 30%–40% who live in industrialized countries are deficient in iron. In the developed world, mineral deficiency is prevented by balanced diets and mineral supplements; however, in developing countries, such countermeasures do not exist and since cereals and legumes are naturally deficient in minerals such as iron and zinc, there is widespread anemia and other diseases caused by mineral deficiency.

Biofortification of staple crops, including increasing the total iron content of edible portions of the plant, decreasing the levels of inhibitors of iron absorption, and increasing the levels of factors that enhance iron absorption through conventional plant breeding strategies or modern methods of biotechnology, are approaches currently under investigation. A study that targets rice as the staple crop has recently been conducted using the conventional plant breeding approach. A 9-month clinical trial using a double-blind dietary intervention in almost 200 religious sisters living in 10 convents around metro Manila, the Philippines, demonstrated that consumption of biofortified iron-rich rice, without any other changes in diet, is efficacious in improving the iron status of women with iron-poor diets.

Genetically engineered *japonica* rice plants have been developed as well; these plants have six times the iron content of their nontransformed rice counterparts. Two plant genes were inserted into the rice genome; one gene encodes the iron transport protein, nicotianamine synthase, and the other encodes ferritin, a storage center for iron. The synergistic action of these two genes allows the rice plant to absorb more iron from the soil, transport it in the plant, and store it in the rice kernel. While the total amount of iron and other minerals in plants is an important determinant of nutritional quality, what really matters is the amount of bioavailable iron and how well it is absorbed by the human gut. To this end, a third gene encoding phytase was also incorporated into this rice line by genetic engineering. Phytase degrades phytate, an anti-nutrient that stores phosphate and binds divalent cations like iron and thus inhibits their absorption in the intestine. Plants have also been developed that express a cysteine-rich metallothionein protein belonging to a family of proteins that are thought to enhance iron absorption from the gut. In the future, high-iron rice such as the examples listed above could help to reduce iron deficiency, especially in developing countries in Africa and Asia.

Calcium-biofortified Crops

Inadequate dietary calcium is a known contributor to osteoporosis. One research group has genetically modified carrots to express increased levels of a plant calcium transporter, by modifying the way the calcium transporter regulates itself. These plants were shown to contain about a twofold-higher calcium content in the edible portions of the carrots. This increase in calcium content was also shown to result in an increase in the total amount of bioavailable calcium in a series of randomized clinical trials, where the amount of calcium incorporated directly into bone could be determined. The results of feeding studies using both mice and humans demonstrated an increase in calcium absorption from the calcium transporter-modified carrots compared with the control carrots. This experiment clearly demonstrates that carrots and other vegetables can be fortified by genetic engineering to carry greatly increased concentrations of bioavailable calcium. It is also possible that the calcium transporter protein can be modified to transport other nutrients, such as zinc, for example, to form additional, much needed biofortified crops.

Recently, tomatoes that have the same calcium-enriched trait were found to have increased fruit firmness and prolonged shelf life, suggesting that calcium-fortified fruits and vegetables may have additional commercial benefits. Furthermore, lettuce expressing the same deregulated calcium transporter protein was shown to contain 25%–32% more calcium than controls. Using a board of highly trained descriptive panelists, biofortified lettuce plants were evaluated and no noteworthy differences were detected in flavor when compared with control plants. Sensory analysis studies are an important factor toward public acceptance of genetically modified foods.

Although this work represents initial studies toward understanding the nutritional impact of transgenic foods, the technology could in theory be applied to other crops because it involves the overexpression of a gene common to all plants. Additionally, the approach in this work can serve as a model system regarding the role of other plant alterations in the bioavailability of nutrients contained in crop plants.

Biofortified Corn with Multivitamins

Is it possible to design a single staple crop, such as rice or corn, that is biofortified to provide every single micronutrient essential to human health and at appropriate levels such that the recommended daily intake of all micronutrients would be achieved with the typical daily consumption of that grain? While perhaps this is not the most attractive method to provide micronutrients, it may be a temporary solution for the most needy. The concept of generating nutritionally complete staple crops could offer an essential short-term solution to global malnutrition, by improving the health of subsistence farmers and providing them with an opportunity to improve their economic status while seeking out a more conventional means of achieving nutritional diversity in their diet. One step toward making this concept a reality was the development of a triple-vitamin fortified corn commonly consumed as a crop plant by rural South Africans. The transgenic kernels contained high amounts of β-carotene, ascorbate, and folate, made possible by genetic modification of several metabolic pathways within the corn plant (Figure 5.10). This achievement

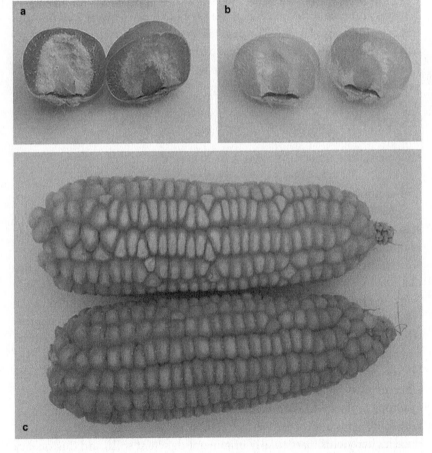

Figure 5.10.
Accumulation of carotenes in the endosperm of transgenic corn line L-1. (A) Orange-yellow phenotype of the transgenic endosperm. (B) Normal phenotype of the WT M37W endosperm. (C) Comparison of WT and transgenic cobs, showing significant increases in the levels of key carotenoids in the transgenic cobs. Photo courtesy of *Proceedings of the National Academy of Sciences USA* 106, no. 19 (May 12, 2009): 7 762–67.

provides the first step toward the development of nutritionally complete cereals to benefit the world's poorest people.

Nutritionally improved cassava provides another example. A quarter of a billion sub-Saharan Africans rely on this starchy tuber as a staple in their diet. Unfortunately, the levels of iron, zinc, and vitamins A and E in cassava are nutritionally insufficient. Projects such as the BioCassava Plus program are working toward developing a superfood version of cassava with increased levels of iron, zinc, protein, and vitamins, as well as pathogen resistance.

In assessing strategies to deal with micronutrient deficiency, the provision of a varied diet with fresh fruit, vegetables, and fish would of course be ideal. However, where this variation in diet is impossible because of poverty and poor governance, super-enhanced, nutritionally complete crops could provide a durable solution to improve the health and general well-being of impoverished populations. The best biofortification strategies may mix genetic engineering with conventional plant breeding. The adoption of biofortified corn will help to improve the health and well-being of the world's poor, but this will be possible only if political differences regarding the use of transgenic crops are resolved.

CAN BIOFORTIFIED FOODS MAKE A DIFFERENCE?

The accumulation of results so far suggests that biofortification can contribute to increased micronutrient intake and improved micronutrient status. There are many benefits to producing biofortified foods for people in developing countries. Biofortification provides a truly feasible means by which nutritionally complete foods can be delivered to malnourished populations in remote rural areas who have limited access to commercially marketed fortified foods. Since these biofortified crops are versions of the regular staples that these people eat on a daily basis, no behavioral changes with regard to family meal planning are required. This is particularly attractive as historically, past interventions in developing nations have depended heavily on supplementation programs, which are sometimes folly to the whim of government initiatives and international funding.

Crops biofortified with higher mineral content tend to be more resistant to diseases and other environmental stresses. These crops also produce higher yields, particularly in mineral-deficient soils, and this quality can be repeated growing season after growing season, and year after year. Thus, biofortified crops are highly sustainable and perhaps even more commercially attractive to rural farmers, once the initial investment in germplasm is made.

There are, of course, some hurdles to be overcome. More research is needed to further identify and characterize the means by which plants control and regulate mineral absorption and uptake from soil to roots, transportation, and distribution to accumulate in different edible tissues. Finally, to be effective, the micronutrient that has accumulated in edible portions of the plant must be bioavailable to people who consume the plant in the form of a regular meal. Just to attempt to determine the latter can be extremely difficult as well as impractical.

Besides this obstacle, other problems could easily arise. For example, biofortified foods are less likely to be accepted by a given population if they look different from their unfortified counterparts. Golden rice and other foods that have been fortified with vitamin A are noticeably darker yellow or orange in color. Vitamin A enhanced maize looks suspiciously different from the white maize eaten by the people of Africa. Yellow maize is negatively associated with animal feed. Similarly, Africans prefer a white-fleshed version of the sweet potato; the orange-fleshed tuber is much less desirable. Furthermore, there are still public concerns over genetic engineering of crops in general; such concerns also exist in Africa and her agricultural trading partners. Issues such as these need to be overcome to assure that biofortified foods will be accepted.

CONCLUSIONS

The identification of chemical and metabolic pathways by which novel functional and biofortified foods are generated has led to a new discipline in plant science that is rapidly gaining momentum. Construction of dietary studies that can truly investigate the effect of single nutrients or food groups within the patterns of a regular diet of an individual remains a challenge.

In industrialized societies, the problem of micronutrient deficiency can be addressed by ensuring that a wide variety of fresh fruits and vegetables are part of a daily diet, along with supplementation and fortification programs to enhance the availability of vitamins and minerals. As many in developing countries would not have such access, and since most of their populations subsist on a monotonous diet of cereal grains that lack essential vitamins and minerals needed for a healthy lifestyle, micronutrient deficiency remains a far-reaching and unfortunate reality. Since people are malnourished, their ability to work effectively is impeded substantially and this fact plays a key role in maintaining the poor socioeconomic conditions that prevail in such regions.

Malnutrition due to micronutrient deficiency remains at a crisis level. Sustainable solutions are required. Successful provision of adequate amounts of micronutrients could dramatically contribute to improving the productivity and livelihood of those who grow crops in nutrient-deficient soils. It would provide substantial health benefits to many people in developing countries, and would contribute greatly to the agricultural and economical development efforts of these countries.

While the best outcome would be to provide a more diversified diet to the world's malnourished through a variety of nutritious crops, this is currently out of the reach of many impoverished people in the developing world. Evidence exists that even small deficiencies in micronutrients can impair the health of children and even increase infant mortality rates. Biofortification, using either conventional plant-breeding strategies (e.g., cross-breeding) or genetic engineering techniques, is one way to enhance the amount and bioavailability of selected nutrients in a staple food item.

Plant breeders are now developing carrots with twice the normal calcium content, tomatoes higher in antioxidants, cassava biofortified with iron, and many other crop types. Several examples of commercially successful biofortified foods are now available. In Australia, zinc-dense wheat varieties have been developed and are already being grown for commercialization, while in the United States an iron-efficient soybean has been developed to flourish in iron-deficient soils. Advances such as these in nutritional and plant sciences are too important to ignore. Food crops with improved micronutrient quality, disease resistance, and bioavailability can make a tangible difference in the lives of many people throughout the world for years to come.

CHAPTER 6

Food Security, Climate Change, and the Future of Farming

Plants as food and medicines are essential components of human health. So far, we have discussed the primary mechanisms and untapped potential of plants to provide medicines for the world's poor, either by identifying active compounds already present in plant species or by using plants as bioreactors to grow and produce new medicines. This book has also argued for the need to develop biofortified foods to stave off the "hidden hunger" of the malnourished, as well as to identify functional foods that have demonstrated the ability to prevent chronic diseases such as cancer. All of these chapters illustrate the potential of plants to provide solutions to substantial problems facing humanity today. This chapter addresses the greatest challenge of all: achieving food security for the next century.

Ensuring that people have an ample supply of nutritious food is a fundamental cornerstone to human health. People who continuously do not get enough to eat find it difficult to recover from even the most preventable of infectious diseases. The immune systems of the chronically hungry are not able to withstand infections that people in developed countries can easily overcome. Today, 1 billion people are undernourished, meaning they do not receive enough calories each day to support their needs.

Clearly, in addition to enhancing the nutritional content of edible crops through biofortification, we will need more food on the whole to keep up with the rapid growth in world population. This will necessitate a new approach to fertilizer usage, including better strategies to prevent environmental degradation due to nitrogen runoff and the overuse of fossil fuels in agricultural production. The looming threat of climate change confounds the global food supply chain even more. Climate change will impact water availability, increase temperatures, and alter growing seasons and the prevalence of pests in ways that cannot easily be foreseen. Furthermore, the danger of channeling much needed food crops into biofuel production puts additional pressure on an already volatile situation.

Resolving these issues will take a radical change in the way agriculture is conducted on many different levels. How will the planet grow crops that can adequately feed an ever burgeoning population, yet with less water and other agricultural inputs available? This chapter will present novel varieties of crops, farming methods, and technologies that will transform our current crop production to one that is sustainable and achievable for the 21st century.

The world population continues to grow and is estimated to reach a plateau of approximately 9.4 billion people by the middle of this century. A subsequent slowing in population growth will be attributed to an increase in wealth, and as a result there will be an even greater demand for meat, processed foods, and other consumables associated with industrialized, developed countries. Along with this population increase, widely fluctuating food prices are predicted to continue, partially as a result of increasing demands for food from rapidly developing countries and increased competition between food crops and first-generation biofuels. Even now, farmers are finding that their current requirements for more land, water, and energy are approaching their limits. These limits, combined with stresses on the environment caused by current agricultural practices—overuse of fertilizers, pesticides, and irrigation—are only further aggrevated by the approach of climate change and its impact on global food production.

Food insecurity goes hand in hand with poverty, with just under 3 billion of the world's people living on less than US $1 per day and another 2 billion only marginally better off. The vast majority of the world's poor are rural farmers in developing countries, subsisting as well as drawing their livings entirely from small-plot farms. Since these farmers are too poor to irrigate their crops or buy fertilizers and pesticides, their soils are exhausted of nutrients and the crops are highly susceptible to pests, diseases, and drought. Rural farmers sometimes feel forced to abandon their land and relocate to cities, where they contribute to the increasing crisis of urban poverty and hunger (Figure 6.1).

The world is predicted to require 70% to 100% more food by 2050. Since the amount of arable land that remains available for agricultural use is limited and may possibly even decrease over the next few years (as a result of increasing urbanization in developing countries as well as the likelihood that currently arable lands may succumb to salination and desertification due to climate change), it becomes increasingly likely that more food will need to be produced from the same or even less acreage. Water will become more and more precious in the next few decades. By far the largest consumer of water is agriculture; it is currently responsible for approximately 70% of total world consumption and is predicted to rise by 35%–60% between 2000 and 2025. Evidently, we must generate new varieties of crops that can produce higher yields using a smaller water supply. Furthermore, since one-third of the world's food is presently produced on irrigated land, changes in weather patterns will also impact water storage and transport networks.

The effects of climate change on world agriculture and food production are far-reaching. Expected to fare the worst is sub-Saharan Africa, with as much as 8% of present arable land with agricultural potential at risk of turning into arid and semi-arid desert by 2080. Long recognized as a drought-prone continent, Africa is likely to see its food production become even more precarious with the advent of climate change.

Climate change is also anticipated to have dire consequences on rice production in Asia. Rice as a crop is particularly vulnerable to high temperatures and prolonged dry seasons. Much of the freshwater available in both China and India is derived from rivers fed by Himalayan glacial meltwater, and by some estimates, these glaciers may lose as much as 80% of their volume by 2035.

Extreme weather conditions brought about by climate change, such as drought and floods, will impact agricultural practices in temperate regions, including those in North America and Europe. Changes in temperature and rainfall patterns will affect the types of crops that can be grown in a particular area and may even alter the surrounding natural habitat, including neighboring wildlife, pests, and pathogens. In order to secure food supplies for the future, significant changes in farming practices will become necessary, counterbalancing increased yield with sustainability of resources.

NORMAN BORLAUG AND THE GREEN REVOLUTION

This not the first time the world has faced the difficult issue of too many mouths to feed. In fact, world famine due to overpopulation was predicted by Thomas Malthus in 1798. The world's population has doubled several times without Malthus's prediction ever actually materializing as a result of the Green Revolution, that is, the introduction of new, high-yielding plant varieties as well as synthetic inputs and more modern irrigation systems. In the1960s, it became clear that India faced extreme food insecurity to the point of mass famine. The pivotal point in India's change of

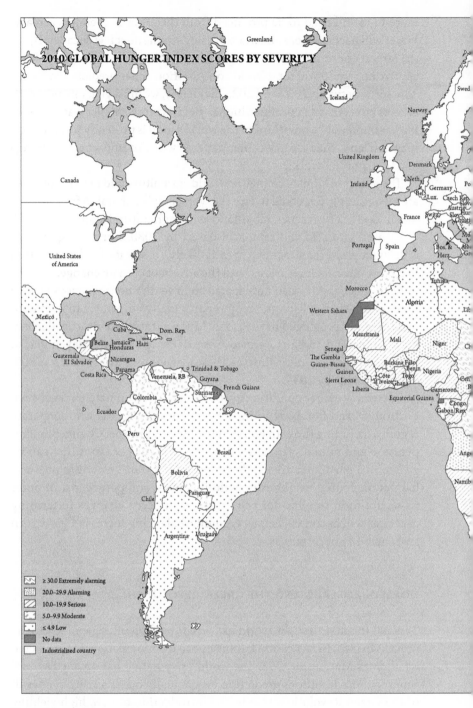

Figure 6.1.
2010 Global Hunger Index scores by severity.

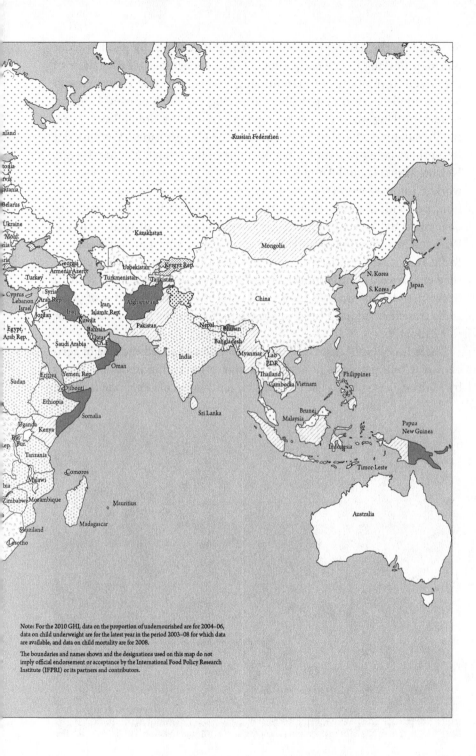

Note: For the 2010 GHI, data on the proportion of undernourished are for 2004–06, data on child underweight are for the latest year in the period 2003–08 for which data are available, and data on child mortality are for 2008.

The boundaries and names shown and the designations used on this map do not imply official endorsement or acceptance by the International Food Policy Research Institute (IFPRI) or its partners and contributors.

fortune was the development of dwarf wheat varieties by plant breeder Norman Borlaug in the early 1960s. Often recognized as the father of the Green Revolution, Norman Borlaug has been considered by many to be the single person responsible for saving the lives of over 1 billion people. Borlaug was one of only six people to win the Nobel Peace Prize, the Presidential Medal of Freedom, and the Congressional Gold Medal (the others being Martin Luther King, Jr., Elie Wiesel, Mother Teresa, Nelson Mandela, and Aung San Suu Kyi); this recognition came for his work to improve world food security. In the presence of adequate irrigation, pesticides, and fertilizers, the new high-yielding wheat varieties developed by Borlaug significantly outperformed traditional varieties. With the introduction of Borlaug's high-yielding wheat, wheat yields for India and other countries nearly doubled in the following quarter century (Figure 6.2). India soon adopted a semi-dwarf rice variety with 10 times the yield of traditional rice. As a result, within a matter of years India transformed itself from a country threatened by famine to a major rice exporter.

There is no doubt that the Green Revolution changed the face of agriculture. Contrary to predictions of the starvation of hundreds of millions by Malthusians such as Paul R. Ehrlich (author of *The Population Bomb*,

Figure 6.2.
(a) Wheat. Photo Copyright © 2007 David Monniaux; (b) Wheat yields in Mexico, India, and Pakistan, 1950 to 2004. Baseline is 500 kg/ha.

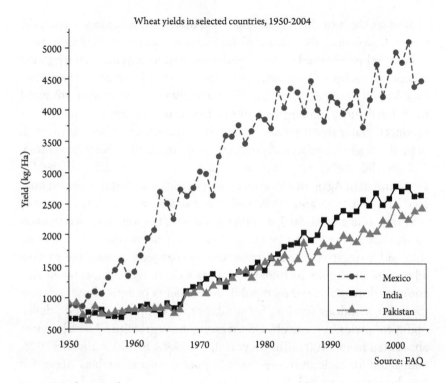

Wheat yields in selected countries, 1950-2004

Source: FAQ

Figure 6.2. (*continued*)

1968), India in fact became self-sustaining in wheat production in 1974 (six years later) by introducing Norman Borlaug's dwarf wheat varieties into their farming practices. This intensification in agriculture brought about by the combined forces of plant breeding for higher yielding crop varieties and modern agricultural practices has also resulted in substantial environmental benefits, such as a greatly lessened requirement for more arable land. For example, between 1970 and 1995, cereal production in Asia has approximately doubled; however, the total land area used for this crop increased by only 4%. Had global cereal yields per hectare not improved, more than 1 billion additional hectares of equivalent land would have been required to match production levels. Acreage such as this was not even available, and if it had been, the environmental consequences due to the deforestation and clearing of land for farming would have been devastating. It was Norman Borlaug who again and again promoted increasing crop yields per acre, thus reducing pressure on natural habitats. This "Borlaug hypothesis"—reducing the need for new farmland, thus preventing the conversion of forests into cropland—ultimately provides a means to preserve the biodiversity of a particular region, a positive gain for wildlife.

Besides the introduction of improved, high-yielding crop varieties, the Green Revolution also extended technologies which at that time were almost exclusively used in developed countries, such as modern irrigation practices, synthetic fertilizers, pesticides, herbicides, and other agrichemicals. An unfortunate consequence is that many agrichemicals are produced from fossil fuels, making agriculture increasingly reliant on petroleum products. Water shortages and health and environmental issues as a result of pesticide use are other end products attributed to the Green Revolution.

How did the Green Revolution lose steam? According to the US Department of Agriculture, wheat yields in the United States almost doubled between 1965 and 2008. Yields of maize rose in a similar manner over the same period. Such a dramatic increase in crop yields led to over-confidence in food security for the West. A general complacency arose that had a negative impact for countries that were never able to take advantage of similar technological advances in agriculture: when rich counties downsize their own public investments in agricultural science, they also withdraw funding for similar types of projects in poor, developing countries. As a result, US investment in agricultural development abroad fell from $400 million a year in the 1980s to $60 million in 2006. Global aid for agricultural research in poor countries such as Africa fell 64% between 1980 and 2003.

More recently, large-scale hunger has been intensified for a number of complex reasons, one of the most important being highly fluctuating food prices, reflecting problems in global economies. Income growth among highly populated countries such as China and the nations in South Asia has led to an increased demand for foods of all sorts, including meat and processed foods, once a hallmark of industrialized countries. The use of crops traditionally eaten as foods as sources for biofuels has also affected the world's food supply.

THE GREEN REVOLUTION MISSED AFRICA

The relative yields of cereal crops in Asia, India, and Mexico have doubled or even tripled over the last 50 years. In contrast, in sub-Saharan Africa, crop yields have in fact stagnated over the past few decades. To this day, as it was 20 years ago, sub-Saharan Africa remains a part of the world where relief from hunger and poverty continues to be elusive. Although agriculture typically makes up 20% to 40% of the gross domestic product, 50% more Africans live below the global poverty line of $1 dollar a day than was the case 20 years ago. For example, in the 1960s, Africa was a net exporter of food, but in the late 1970s it imported 4.4 million tons of food a year; by the mid-1980s, this was 10 million tons, and by 2002, 19 million tons of food grain were imported into sub-Saharan Africa.

Approximately one-third of Africans are undernourished and more than half are malnourished, lacking the essential nutrients and vitamins required for a healthy life. As with the rest of the world, agricultural development in Africa will be significantly impacted by climate change. Extreme heat and desertification are two of the challenges that crops grown in Africa will soon face. Furthermore, in the next few decades, Africa as a continent faces a doubling of the current population to approximately 2 billion people. Much of this surge in population will be focused in cities and major urban centers, so that urbanization will play a much greater role in future land management. Therefore, as if the plight of Africans could not get much worse, the actual amount of arable land available for African farmers is predicted to decrease due to the combined effects of climate change and urbanization.

Farming status in Africa remains grim. Even without the combined pressures of population growth and climate change, Africa needs passable roads, modern irrigation systems, and better quality soils to ensure that it can feed itself. Many smallholder farmers in Africa still work with hand hoes or crude wooden plows (Figure 6.3). Farm animals used for meat, milk, and draft power are stunted by poor health. Agricultural workers continue to use traditional farming practices that are impractical, as well as low-yielding plant seeds that have not been improved by plant breeding. All of these factors constrain the productivity of African farmers and bind them to a continuous cycle of poverty. Two-thirds of Africans are poor farmers in desperate need of new technologies that will help to boost productivity of their own crops. These farmers may not have access to the technical knowledge and skills required to increase production and they

(a)   (b)

Figure 6.3.
Farmers from two parts of the world confront insect pests. (a) Photo from US Department of Agriculture; (b) Africa Renewal, United Nations. Photo by Christine Nesbitt, for IFAD.

lack the financial power to invest in higher production accessories such as fertilizer, machinery, and irrigation systems or the variety of crops that maximize yields. After harvest or slaughter, they may not be able to store the produce or have access to the infrastructure to transport the produce to consumer markets. Other economic conditions brought about by the global market may also impact African food producers.

This sort of deep rural poverty was a constant way of life in both Europe and North America earlier, and it was relieved only with scientific improvements ranging from labor-saving mechanical technologies to better seeds. While modernizing farming practices would be of enormous benefit to Africa's farmers, doing so will be difficult in the near future. Certainly, improved, high-yielding crops could offer great advantages. Better crops are the results of both plant breeding and biotechnological strategies, including genetic modification.

At the time the first genetically modified crops were entering the marketplace, Europe and Great Britain were hit with the previously unheard of prion disease known as bovine spongiform encephalopathy, or mad cow disease. At that time, the public worried whether meat from beef cattle in the United Kingdom was in fact safe to eat. Europeans became mistrustful of any new technology, such as genetically modified (GM) foods. Although they assisted farmers in insect and weed control, the first generation of GM crops in the West had no apparent benefit to food consumers. They didn't necessarily cost less, taste better, or add nutritional value. Even so, poor countries like Africa had much to gain from GM foods, but African governments, out of reverence for the European example, have driven GM crops out of their own market by adapting European-style regulatory approaches toward the new technology. Africa's urban political leadership class still feels a strong international dependence on Europe. As a result, foreign assistance programs and nongovernmental organizations (NGOs) operated and funded by European countries have played an integral role in encouraging Africa to reject modern agricultural science.

A consequence of resistance to GM technologies is illustrated in the unfortunate example of 2002, when Zambia turned down a 10,000-ton shipment of GM corn in spite of its 3 million hungry people, even though this same country had been importing it for several years. Zambian president Levy Mwanawasa commented that simply because his people were hungry that their hunger was no justification "to give them poison"—that is, to give them food that is intrinsically dangerous to their health. The World Food Programme had to remove the GM food aid. In one town, hungry villagers mobbed an armed guard and looted the GM maize before it could be removed. This was the same corn that had been grown and eaten in the United States and Canada since 1996. As Michael Spector writes in his book *Denialism*, "In the end it is not Africa's science-starved farmers that are rejecting biotechnology; it is Africa's urbanized governing elites that are doing so, prompted by their continued deference to urbanized elites in Europe."[1]

AGRICULTURAL MAKEOVER: SUSTAINABLE INTENSIFICATION

It is estimated that a 50% increase in grain yield will be required over the next 50 years to ensure food security for all. Exactly how can food production be increased so that the world's poor are not left hungry? Instead of employing a "one size fits all" tactic, strategies to increase the world's breadbasket must become highly specific to a particular agricultural setting. Tailored approaches must address problems in the food production pipeline in any given region and situation, be it pest management, minimization of waste, or even transport to market. To improve food security, then, new technologies must be developed to accelerate crop improvement through plant breeding, and these must be both economically accessible and readily disseminated.

This radical new way of thinking about agricultural production, of producing more food from the same area of land while reducing environmental impacts, is often referred to as "sustainable intensification." In terms of plant varieties themselves, it involves the use and improvement of conventional and molecular breeding, as well as molecular genetic modification to adapt our existing food crops to increasing temperatures, decreased water availability in some places and flooding in others, rising salinity, and changing pathogen and insect threats. In addition, crops with lower fertilizer requirements are urgently needed.

Advances in plant transformation and gene expression technologies will play a major role in increasing crop productivity and nutritional value. Genetically modified crops can be designed that combat hunger by making plants simultaneously more nutrient rich and less prone to disease and other biotic stresses. The end result of boosting productivity of crops through biotechnology is that food can become affordable to the world's poorest. Biotechnology can pull people in developing countries out of a continuous cycle of poverty. Through agricultural biotechnology, soils that are now of only marginal use will in the future be able to support new, tougher crop varieties that are better able to extract micronutrients and minerals from the soil as well as withstand more extreme environmental conditions such as floods or drought. Soils not fit for use can be regenerated by plant varieties that have been developed for phytoremediation. Crop plants with modified architecture will be sturdier and produce more fruit, tubers, or seeds. New plant varieties will possess alterations in their photosynthetic pathways or modifications in their abilities to metabolize carbon and nitrogen in order to withstand new microclimates resulting from climate change. These tactics for crop improvement, in combination with modifications in water use and soil quality, can help attain the goal of producing enough food to cope with the anticipated increase in population.

UP–AND-COMING TECHNOLOGIES: CROP IMPROVEMENT

A wide variety of agricultural technologies aimed at crop improvement are under development. Here are a few examples.

Herbicide Tolerance

Improvement of crop yields will require the development of plants that are tolerant to herbicides. The herbicide glyphosate—N-(phosphonomethyl) glycine—has been the dominant herbicide worldwide since it was introduced to the commercial market in 1974. Glyphosate's mode of action is to block an enzyme involved in the synthesis of the amino acids tyrosine, tryptophan, and phenylalanine, building blocks used in protein synthesis. Another important gene in the history of herbicide resistance is the herbicide bialaphos (*bar*) gene. The *bar* gene was originally cloned from the bacterium *Streptomyces hygroscopicus*. Since bialaphos also inhibits glutamine synthetase, the *bar* gene has been used to create herbicide-resistant crops.

Transgenic crops that are tolerant to the popular herbicide glyphosate are routinely grown in many parts of the world. Herbicide-tolerant plants have also been generated without using genetic modification. For example, natural glyphosate herbicide resistance genes have been identified in wheat germplasm; from these, individual herbicide-resistant plants have been generated. Taking this a step further, the genes that control herbicide resistance can be isolated and used to generate glyphosate herbicide resistance in plants that lack natural herbicide resistance genes.

Crops Resistant to Drought and High-Saline Conditions

Another area in which agricultural biotechnology is rapidly advancing due to intense research is in the development of crops resistant to environmental stresses. Due to growing concerns regarding climate change, efforts are being made to produce plants resistant to environmental conditions that can be both adverse and extreme, such as drought and flood, excessive temperatures, and high salt levels. There is a continuing need to develop novel plant varieties that are less susceptible to damage or loss by such stresses. Dehydration, for example, a major form of osmotic stress in cells, has been addressed by expressing stress tolerance genes from different organisms to engineer drought tolerance in crop plants (Figure 6.4).

Besides improving biochemical pathways inherent in the plant to resist environmental stresses, water usage can also be addressed by focusing on the architecture of the plant itself. By redesigning root and leaf structure, plants can be generated to have increased efficiency in their ability to uptake and release water. Roots designed for shallow growth in soil, for instance, can better tap soil-surface moisture.

One of the most severe environmental stresses that crops may contend with is high salinity or salt content in soils. High salinity can affect at least 20% of irrigated land worldwide and exerts its negative impact mainly by disrupting the ionic and osmotic equilibrium of the cell. In saline soils, high levels of sodium ions lead to plant growth inhibition and, in certain circumstances, death. Many plants exposed to such conditions have the ability to upregulate various gene products in response to high salinity stress which then cross-talk with each other and act in a coordinated fashion to confer salinity tolerance. Mechanisms of salinity tolerance that plants exhibit can range from sequestering sodium within the vacuoles of their cells via an antiporter gene, to blocking entry of sodium ions into the cell, to even increasing the efficiency of water conservation through the regulation of stomatal closure responses. Mechanisms such as these have been

Figure 6.4.
Drought-resistant rice. From the International Rice Research Institute.

identified in some plant species, and the genes involved can be transferred to other species, such as rice, to make them more salt tolerant. Such plants can then be grown in areas that were previously unsuitable for growth.

Reduced Fertilizer Use

In the context of global environmental change, the efficiency of nitrogen use has also emerged as a key target. Human activity has already more than doubled the amount of atmospheric nitrogen that becomes fixed annually, and this has led to environmental impacts, such as increased water pollution and the emission of greenhouse gases. In many countries, nitrogen inputs are now under the control of legislation that limits fertilizer use in agriculture. Fertilizers are now commonly the highest input cost for farmers, due to rising energy costs. New crop varieties will need to be more efficient in their use of reduced nitrogen than current varieties are. Therefore, it is important that breeding programs develop strategies that select for yield and quality with lower nitrogen inputs.

The requirement for nitrogen-based fertilizers for crops can be addressed in other ways as well. Researchers have developed a rice plant that expresses a nitrogen utilization protein, known as alanine aminotransferase, in its roots. Since these plants have improved nitrogen uptake, they can reach greater yields without the need for high levels of nitrogen from fertilizers. In addition, an emerging area of engineering known as synthetic biology is being applied to design a system that allows nitrogen-fixing bacteria to interact with the root systems of crop plants, such as corn. The bacteria would then colonize the roots and fix nitrogen for the plants, as they naturally do for soybeans. Successfully producing corn and other crop plants that could fix their own nitrogen would greatly reduce the need for chemical fertilizers and be a major step toward sustainable agriculture.

Other means of protecting against environmental stresses are also being considered. For example, crops could be generated that contain a pigment gene that causes a color change in the leaves or stems in times of stress. Crops expressing this gene could provide a warning signal to help farmers respond quickly to pressures such as drought.

Disease-Resistant Crops

One of the first disease-resistant crops developed for commercial use expresses the Bt (*Bacillus thuringiensis*) toxin gene. When an insect susceptible to this toxin ingests the transgenic crop cultivar expressing the Bt

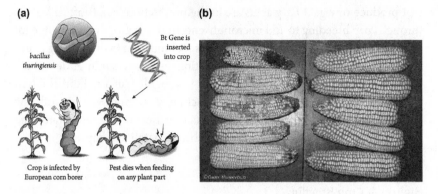

(a)

bacillus
thuringiensis

Bt Gene is
inserted
into crop

Crop is infected by
European corn borer

Pest dies when feeding
on any plant part

(b)

©GARY MUNKVOLD

Figure 6.5.
(a) Bt mechanism of action. Art by Jiang Long and Jen Philpot; (b) European corn borer damage and fungal infection in non-Bt (left) and Bt hybrids following manual infestation with second-generation corn borer larvae.
Source: scq.ubc.ca.
Source: G.P. Munkvold, Iowa State University.

protein, it dies shortly afterward (Figure 6.5). Plants can be engineered to express this biological pesticide, thereby ending the need for external application of harmful chemicals. By expressing the Bt toxin within plant cells, transgenic plants are able to avoid damage from these pests. Crop plants have now been engineered to utilize a number of novel techniques that provide resistance against a variety of pathogens, including viruses, bacteria, fungi, and nematodes.

Next-Generation Plants

Plant scientists are moving toward the development of plants that express more complex physiological pathways involving the combinatorial effect of several plant genes at once. For example, the genes necessary to carry out photosynthesis have been examined and altered to help the plant survive changing weather patterns. Vast changes in genetic makeup of plants may require the design of artificial minichromosomes, in which suites of genes are added to a DNA scaffold. These minichromosomes, introduced into current crop varieties, will maintain plants under conditions that would scorch their native counterparts.

Another rapidly expanding field in agriculture that indirectly addresses climate change concerns the development of biofortified foods, including crops whose edible tissues possess enhanced nutrient content. Scientists have produced world grain staples such as rice, wheat, and maize that are biofortified in vitamins and minerals, and oilseed crops such as soybeans that produce omega-3 fatty acids are in the production line. Biofortification through crop breeding to add micronutrients to staple crops may become increasingly important as a means to address malnutrition in developing countries. The World Health Organization estimates that vitamin A deficiency alone is responsible for causing blindness in roughly half a million children each year. Increased levels of nutrients including vitamin A, iron, and zinc have already been developed in staples such as rice and sweet potato. The generation of biofortified crops was described in detail in Chapter 5.

High-Tech Crop Breeding

With the oncoming combined crises of world hunger and climate change, crop varieties that are both nutritious and capable of withstanding more extreme environmental pressures are required without delay. A deeper understanding of plant physiology, combined with modern recombinant DNA techniques, has enabled scientists to identify new plant varieties that will be more capable of withstanding exposure to adverse environmental conditions, based upon crops currently used as staple foods in developing countries. Often the gene pool of a domestic crop is smaller than that of its wild relatives due to losses of genetic material during generations of breeding. By scanning the genome of wild plants, these lost attributes could be replaced and contribute to the overall fitness of a particular crop so that it can flourish in a hostile environment.

Historically, plant breeding has been a long and meticulous practice. One approach that has been taken to hasten the process is the development of robotic breeding and greenhouse systems. Automated greenhouses now house plant seedlings that are systematically exposed to an assortment of controlled environmental conditions such as heat and drought. The seedlings, which are moved from one condition to the next on conveyor belts, can be monitored in a noninvasive manner entirely by robotics. This process enables crop scientists to identify improved plant varieties with the desired traits, without the need for lengthy field testing. In some instances, the selection of seedlings and the field trials themselves are being handled by computer programs.

New genomic techniques, such as marker-assisted breeding, provide for greater selectivity and reduce the inherent risks found in plant breeding. These techniques are now being employed to identify new crop qualities ranging from submergence tolerance to increased pest resistance. Other novel breeding technologies include marker-assisted selection (MAS), a process by which specific regions of a plant's genome are followed during the breeding cycle. In this way, the trait in question may be tracked regardless of whether it is phenotypically recessive or has a low probability of inheritance. Marker-assisted selection can also be used for gene stacking, a process by which multiple traits such as two or more genes conferring pathogen resistance can be chosen and selected for.

IMPROVED FARMING TECHNIQUES

Besides improving crop varieties themselves, improvements in farming methods can also significantly improve food production. Some of these highlights are discussed next.

Precision Agriculture

Crop productivity can be increased at the same time that sustainability is maintained and potential negative environmental impacts are reduced through a new discipline known as precision agriculture. Precision agriculture refers to an information-based form of management for agricultural production systems that strives to optimize resources currently in place. This can entail the utilization of sophisticated devices such as remote sensing and global navigation satellite systems to illustrate in detail any variations in soil conditions, changes in moisture or nutrient requirements, or even the presence of pests or weeds, and to respond to these variations in an appropriate fashion so as to enhance crop growth (Figure 6.6). The ability to monitor highly localized environmental conditions will enable agriculturalists to determine whether crops are growing at maximum efficiency or to identify a problem and its exact location. In this way, water, nutrients, and pesticides can be applied only to the places and at the times they are required, thereby optimizing the use of inputs and avoiding unnecessary waste of water, nitrogen runoff, and overuse of pesticides. Through precision technology, farmers no longer need to treat a field of crops as a single homogeneous unit. For example, just as it is now possible for farmers to utilize a global positioning system (GPS) system to monitor their tractor's location to within roughly a one-meter radius, it will soon become feasible for them to assess a series of geographical information system maps that show which specific areas within their fields require moisture, where pests may be present, and where more nutrients may be needed. Precision farmers can monitor variables that affect their crops, including soil moisture, surface temperature, photosynthetic activity, and weed or pest infestations, by measuring precisely the way their fields reflect and emit energy at visible and infrared wavelengths. These

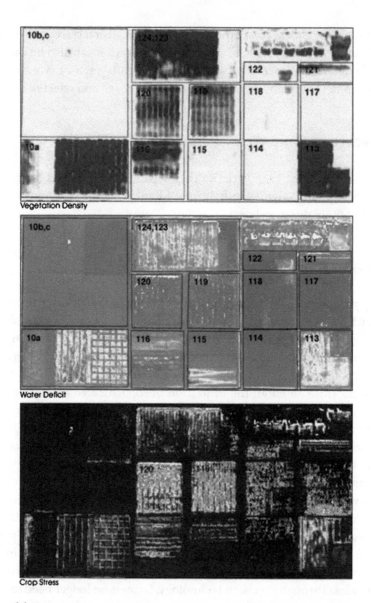

Figure 6.6.
False-color images demonstrate remote sensing applications in precision farming. These three false-color images demonstrate some of the applications of remote sensing in precision farming. The goal of precision farming is to improve farmers' profits and harvest yields while reducing the negative impacts of farming on the environment that come from over-application of chemicals. The images were acquired by the Daedalus sensor aboard a NASA aircraft flying over the Maricopa Agricultural Center in Arizona. The top image shows the color variations determined by crop density (also referred to as #147 "Normalized Difference Vegetation Index," or NDVI), where dark blues and greens indicate lush vegetation and reds show areas of bare soil. The middle image is a map of water deficit, derived from the Daedalus reflectance and temperature measurements. Greens and blues indicate wet soil and reds are dry soil. The bottom image shows where crops are under serious stress, as is particularly the case in Fields 120 and 119 (indicated by red and yellow pixels). These fields were due to be irrigated the following day. Courtesy NASA Earth Observatory.

data will be uploaded to a "variable rate technology device," an instrument added to their tractors that processes information gathered from remote sensors and then regulates exactly how much water, fertilizer, and pesticide need to be applied at precisely the right location in the field. The objective of this manner of farming is to simultaneously increase crop yield while reducing as much impact on the environment through chemical overuse as possible. Through the continuous monitoring of any environmental variables and by subsequently applying water, fertilizer, and pesticides or herbicides in a highly fine-tuned, targeted fashion, crop production can be optimized while the amount of chemical runoff and overuse of water is reduced significantly. Precision agriculture, therefore, can greatly improve the food supply by managing the quality of agricultural produce, while at the same time promoting sustainability of the environment.

Precision agriculture technology can be taken a step further and provide, via computer data exchange, a well-marked trail of any plant food produce from the time it is grown as a crop in the field to the time it is sold in the supermarket. Theoretically, these improvements in information technology will make it possible to take the food chain infrastructure map back to the production of a food plant at its precise place of growth in the field. It will also help to determine in a timely fashion how much more of a crop should be produced, thus ensuring proper supply and quality control—with a minimum of waste.

How can the sophisticated principles of precision agriculture such as satellite-directed remote sensing be applied to poor rural farmers in developing countries? One successful approach has been the use of low-tech strategies such as drip irrigation for drought-prone areas. Also known as microirrigation, drip irrigation conserves both water and fertilizer by enabling water to slowly drip through a network of pipes to the roots of plants (Figure 6.7). In the same vein, a number of developing countries have adapted the technique of micro-dosing the base of some crop plants with minute amounts of fertilizers at critical points in the plants' growth cycle, such as at the time of sowing and later on, when the plants are growing rapidly, to provide plants with the nutrients they need at optimal times of their development. This low-tech approach has been demonstrated to increase yields substantially and would be beneficial to countries that have little access to artificial fertilizers. For example, for the past 30 years or more, China has been the largest user of synthetic fertilizers. While it has helped farmers nearly double their crop productivity, the overuse of fertilizers in China has brought environmental issues such as the accumulation of nitrates, resulting in increased greenhouse gas emissions and pollution issues. Soil scientists are recommending new farming techniques that use much less fertilizer while maintaining or even increasing crop yield.

Figure 6.7.
Drip irrigation system. Photo by "iDE/Bimala Colavito." Courtesy of iDE.org.

Nanotech and Beyond

Nanotechnology, or the study of technologies based on controlling matter at an atomic or molecular scale, is being positioned to transform and revolutionize the entire food industry from the way food is produced to the way it is packaged and transported.

For example, nanosensors, autonomous sensors linked into a GPS system for real-time monitoring, can be distributed throughout the field where they can gauge soil conditions and crop growth. Alternatively, nanosensors can assist grocery store personnel to identify any food items that have passed their expiration date.

Besides its importance in the development of smart sensors, nanotechnology is also used in the form of catalysts and delivery systems that will help agricultural workers detect and combat plant pathogens more efficiently, through encapsulation and controlled release techniques. Nanotechnologies will revolutionize the use of pesticides and herbicides. In some products, nanocapsules can remain inert until they contact leaves or insect digestive tracts, at which point they are designed to release a pesticide. For example, the nanocapsule "gutbuster" (made by Syngenta) breaks open to release its contents only when it comes into contact with alkaline environments, such as the stomach lining of certain insects. Along these lines, nanoparticles are being developed that will release their cargo in a controlled manner in response to different signals, such as heat or moisture.

Alternatively, pesticides and herbicides can be applied more easily and safely in the form of nanoemulsions, which can be either water or oil-based. Nanoemulsions can be easily incorporated into the form of gels and creams, and have multiple applications. For example, nanoemulsions containing plant growth regulators can be applied to crops early in their growth season and strengthen their general constitution at the onset of various stresses such as heat and drought.

Nanoparticles are even finding their way into novel technologies to deal with the effects of past agrichemical overuse. For example, lanthanum nanoparticles that are now available can absorb phosphates from aqueous environments. Applying these in ponds and swimming pools effectively removes available phosphates and as a result prevents the growth of algae, thus offering a solution to the "algal blooms" associated with nitrogen runoff from fertilized fields.

Reducing Food Waste

Whether a country is highly industrialized or not, approximately 30% to 40% of its food ends up as waste. In wealthier societies where produce is plentiful in supermarkets, excess waste can be managed in a more efficient fashion. In the developing world, food waste is often the result of a poor farm-to-consumer food chain infrastructure or inadequate storage facilities, with unrefrigerated food spoiling or succumbing to contamination by pests before it ever reaches the consumer. Farmers often must carry their harvests down primitive dirt paths to other towns and villages, and therefore find it difficult to transport their harvests to distant markets—which may take days to reach by foot. By controlling ripening and senescence of fruits and vegetables through the use of biotechnology, spoilage could be reduced in crops such as tomatoes and bananas with enhanced shelf lives. If issues such as food transport, distribution, and storage could be addressed, the total amount of food available for consumers in these countries would increase dramatically.

ENVIRONMENTAL USES FOR PLANTS

Plants are also being developed to improve human lives at a less direct level. One of the most promising and controversial agricultural technologies involves the use of crops for biofuel. Plants have also been developed for cleaning up man-made pollutants in the environment and thus increasing the amount of arable land available for farming without encroaching upon current land preserved for wildlife. These technologies are addressed in the next section.

At the moment, there is much controversy over the topic of biofuels. Plant-based liquid fuels such as ethanol could potentially act as a "greener" alternative to fossil fuels and help to break foreign oil dependence. Conversely, biofuels are also believed by many to be responsible for driving up food prices by competing with food crops for valuable prime farmland and for causing environmental damage.

Coming to the forefront, however, is the development of biofuels from nonfood crops, grown on less than optimum farmland. One leading candidate in this field is switchgrass, a plant that appears to grow with next to no effort throughout the prairies of the US Midwest. Instead of taking away a food crop to create fuel, as is the case with corn, for example, switchgrass cellulosic ethanol promises fuel from a plant that is considered to be more of a weed than anything else (Figure 6.8). One great advantage of using switchgrass as an energy crop is that it thrives on what is considered marginal land, acreage not currently needed for corn and

Figure 6.8.
Switchgrass (*Panicum virgatum*).

other food crops. A long-term, large-scale field study involving fields of farms consisting of 15–20 acres in the US states of Nebraska and North and South Dakota showed that switchgrass yielded 540% more energy as a biofuel than the amount of energy used to grow, harvest, and process it. In addition, greenhouse gas emissions from switchgrass fuel would be 94% lower than emissions from petroleum fuel.

Besides switchgrass, all sorts of plant material are being assessed for their potential value as biofuels. For example, since the ethanol produced for biofuel comes from cellulose, transgenic plants that express altered or improved cellulose or other biopathways which could be more conducive to greater levels of biofuel production are under production. Biofuel is also produced from organisms as diverse as algae and fungi, none of which would impact the world's food supply in the way that food crops such as corn would. It only takes a quick glance at the patent literature to see the large scope and range in the potential for plants to enhance energy security, without undermining food security at the same time.

Restoring the Fertility of Land by Phytoremediation

The presence of pollutants such as heavy metals in soil and water have been a long-standing issue due to their persistence in the environment and their potential to act as carcinogens to humans and wildlife. In a process known as phytoremediation, various plant species can decontaminate soil or water by uptaking heavy metals through their root systems and translocating them into various tissues such as foliage where they can accumulate and be

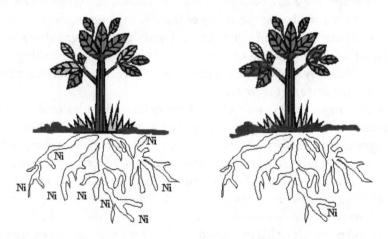

Figure 6.9.
Typical In situ Phytoremediation System.
Source: Federal Remediation Technologies Roundtable (frtr.gov).

stored without causing further damage. The metal ion accumulated in these tissues can then be removed and disposed of or burned to recover the heavy metal. Over 400 plant species have been identified as potential candidates for soil and water remediation (Figure 6.9). Plants, along with associated microorganisms, can also be used to degrade organic pollutants. Even common pollutants such as phosphorous can be phytoremediated using plants. Removing heavy metals and pollutants from the environment can increase the amount of arable land available for food production.

AGRICULTURAL SUSTAINABILITY AND ORGANIC FOOD

Could organic farming also be considered a component of sustainable intensification? The next section explores this as a possibility.

The term "organic farming" was first used in 1939 by the British agriculturalist and writer Lord Northbourne, who is often acclaimed as the "father" of organic agriculture. Today in the United States alone, organic agriculture has become a $23 billion business. Organic foods now are considered by many to constitute some sort of ethical superiority, and consumers, believing that food produced under organic farming practices improves their health or is better to the earth, are more than willing to pay much more for organic food than for conventional food. Indeed, part of this new resolve for consumers can be explained as a response to increased reports of health problems in industrialized countries and the relationship between poor eating habits, junk food, and obesity. For some, organic food also represents an imagined simpler time when small, family-run farms producing a diverse combination of crops and livestock were the norm in the rural landscape. Many of those who purchase organic products cultivate an image of a mythical and nostalgic rural agriculture, reminiscent of a time when life was less complicated and farming was accomplished by simple farmers, not in the highly industrialized production of agribusiness (Figure 6.10).

One danger in this way of thinking is that this idealized view of farming in the past is highly unrealistic. Any glimpse into the history of agriculture in North America, for example, provides a more grim and accurate picture of rural people, many in poverty and malnourished, trying to make a meager living off the land. It was in fact the upgrading of agricultural practices, including the use of modern farm machinery, irrigation systems, new crop cultivars, and synthetic fertilizers and pesticides, that improved yield and made crops less expensive to grow. This in turn lowered food prices, freed workers from backbreaking labor, and allowed people to improve their incomes—and

Figure 6.10.
American Farm Scene, engraved by Nathaniel Currier (1813–98) after a painting by Frances (Fanny) Flora Bond Palmer (c.1812–76). Pub. by Currier and Ives, New York (colour litho). Private Collection/ Giraudon/ The Bridgeman Art Library.

eventually helped America to become the economic giant that it is today. Throughout the history of civilization, advances in agriculture have been intrinsic to a country's economic growth and, with only a few exceptions, a step toward the development of wealth.

Nonetheless, this nostalgic and sentimental view remains a popular fantasy. Those who reside in wealthy countries have developed a negative impression of much of modern agriculture, including the use of pesticides, fertilizers, and, to an extent, genetically modified crops. While these technologies were critical in making farmland in developed countries more productive, they also brought with them an industrialization of agriculture that remains unappealing to many. The organic food system, however, has morphed into two different production and sale operation models: the small local one (farmers' markets and local cooperative community centers) and the large industrial one (supermarkets). Nowadays, organic food is available for sale even in major superstores such as Wal-Mart. In many respects, the organic movement has become indistinguishable from big business. Unbenown to some, this large-scale industrialization of organic food often means that smaller organic farms have been bought out by a small number of giant corporate organic growers that now dominate the market for fresh organic produce sold in the United States. Indeed, much of the organic crops sold in the United States are grown by the same conglomerates that sell conventional foods. As a result, the smaller organic farmers find it impossible to compete with these large companies and have trouble staying afloat on the wholesale market. While it may not have been the initial aim of organic farming to move in this fashion, it is simply more cost-effective for a grocery supermarket chain to buy from a single large industrial organic farm than to make multiple purchases from many smaller farms.

When it comes to discussing industrialized foods, no one does it better than Michael Pollan. In his book *The Omnivore's Dilemma*, Pollan states that organic philosophy and practices "have been stretched and twisted to admit the very sort of industrial practices for which it once offered a critique and an alternative."[2] In his term, the "supermarket pastoral," "people have been seduced to believe in a utopian past in which farm animals live as they did in books we read as children."[3] In this world, there is a widespread belief in safe food, a connection to the earth, and a "back to nature" philosophy. Industrial organic, as Pollan puts it, looks to many as a contradiction in terms. "The organic movement, as it was once called, has come a remarkably long way in the last thirty years, to the point where it looks considerably less like a movement than a big business."[4] Pamela Ronald and her husband Raoul Adamchak concur with this concept in their book *Tomorrow's Table*: "This dilution of meanings and practice or organic agriculture-lucrative segments of organic commodity chains are being appropriated by agribusiness firms, many of which are abandoning the agronomic and marketing practices associated with organic agriculture."[5] Pollan has a very good point when he laments, "Can a box of salad that was packed into a plastic container filled with inert gas (to increase longevity) 5 days ago and has traveled over 3,000 miles truly be organic?"[6]

Many hold the belief that organically grown crops are better for one's health and for the environment. However, a 12-month systematic review commissioned by the Farm Service Agency carried out by the London School of Hygiene and Tropical Medicine based on 50 years' worth of collected evidence concluded that "there is no good evidence that consumption of organic food is beneficial to health in relation to nutrient content." Other studies have found no proof that organic food offers greater nutritional value, more consumer safety, or any distinguishable difference in taste.

Consumers often cite concerns about pesticides as being a reason to "go organic," despite statements from both the Mayo Clinic and the American Cancer Society about the harmlessness of the low levels of residual pesticides on conventional food. In fact, Ronald Paarlberg, in his book *Food Politics*, explains:

The United Nations, through the Food and Agricultural Organization (FAO), and the World Health Organization (WHO), has established acceptable daily intake (ADI) levels for each separate pesticide. The ADI level is set conservatively at 1/100 of an exposure level that still does not cause toxicity in laboratory animals. Moreover, actual residue levels in the United States on conventional foods are well below the ADI level. For example, when FDA surveyed the highest exposures of 38 chemicals in the diets of various population subgroups, it found that for 4 of these 38 chemicals, estimated exposures were less than 5 per cent of the ADI level. For the other 34 chemicals, estimated exposures were even lower, at less than 1 percent of the ADI level.[7]

There is no reason to believe that "natural" chemicals used in organic food production are any safer than synthetic pesticides although some propose that utilization of organic fertilizers would be a more environmentally friendly approach to farming. Much of the world uses synthetic fertilizers which provide nitrogen, an essential building block required for plant growth, and changing this is unrealistic. Robert Paarlberg offers the following scenario: "To replace that synthetic nitrogen with organic nitrogen would require the manure production of approximately 7–8 billion additional cattle, roughly a five-fold increase from the current number of 1.3 billion. The United States alone would have to accept another 1 billion additional animals and an added 2 billion acres of forage crops to feed those animals, equal to all the land in America except Alaska."[8]

Other environmental concerns also exist. For large, organic industrial farms, heavy tillage in the field destroys the tilth of the soil and reduces its biological activity, releasing nitrogen in the air and disturbing nitrogen-fixing bacteria so that even more organic fertilizer is required. Organic crops sold by supermarkets often were grown in large, industrial farms thousands of miles away, and transporting them to market further increases the carbon footprint involved in making the product.

One of the principal criticisms from the scientific community concerning organic farming is its lower crop yield when compared to conventionally grown plants, and even more so when compared to GM plants. Lower crop yield leads to more wilderness plowed for farm use, which again leads to environmental concerns such as a loss of wildlife, habitat, and biodiversity. There is simply not enough arable land on the planet to feed the world's burgeoning population as it stands today, using organic farming practices. Moreover, if the number of lower yielding organic farms are expanded, land that has currently been set aside for wildlife will be forced into farmland.

Meanwhile, as the arguments for or against organic versus conventional food rage on, the other several billion or so people on the planet have had organic farming imposed upon them. They cannot afford synthetic pesticides or fertilizers; have no access to new, improved, disease-resistant varieties of crops; farm on nutrient-poor soil; and are less successful with their crops as each year goes by. The organic movement of the wealthy countries has relegated African farmers to lives of increased poverty and misery. It may be true that for some who are skilled in organic farming practices and who work in regions containing fertile soil, barring an unexpected infestation by insects or other pathogens, some organic crop yields may resemble those of conventionally grown crops. This is not likely to ever be the case in nutrient-deprived sub-Saharan Africa. As Michael Specter put it in his book *Denialism*, "Insisting that the world can feed nine billion people with organic food is nothing more than utopian extremism. An organic universe sounds delightful, but would consign millions of people in Africa and much of Asia to malnutrition and death."[9]

The organic movement has been called a romantic ideal. It believes it is environmentally friendly, but it avoids rigorous scientific criticism. The environmentally conscious must look at the data, and those data do not support the growth of organic farming as a way to feed the world. While some may consider organic food to be simply another harmless fad, so-called politically correct worldviews such as these can be detrimental to the food insecure, such as sub-Saharan Africans. Norman Borlaug once said, "While the affluent nations can certainly afford to adopt ultra low-risk positions, and pay more for food produced by the so-called 'organic' methods, the one billion chronically undernourished people of the low income, food-deficit nations cannot."[10]

CONCLUSIONS

The world faces a daunting challenge. Within the next 20–30 years, we must find a way to produce more food, most likely with less arable land and water, and with fewer fertilizers and pesticides than we use today. In addition, we must abruptly curb the level of greenhouse gases that are being emitted while at the same time cope with those consequences of climate change that are now irreversible. To accomplish this, new technologies and farming practices must be introduced and implemented as soon as possible.

One of the major solutions to global food security lies in the marriage of biotechnology and plant breeding strategies. For example, the biofortification of key food crops with a higher availability of micronutrients such as iron and vitamin A is now possible. Other strategies to address food insecurity must include the generation of plants that produce higher yields, either by increasing the intrinsic yield capability of crop plants or by protecting them from drought, high temperature, and diseases. Enhanced crops will be able to grow in places that are currently unsuitable and can be developed as a result of genome sequencing and identification of new genetic traits in weedy relatives or in plants that now thrive in such conditions. A significant proportion of this genetic potential can be realized in food crops through biotechnology. Crops will be developed that will photosynthesize more efficiently while requiring less water and fertilizers, thus ensuring a higher yield with a lower input than is common in agricultural practices today. Through biotechnology and selective plant breeding, sustainable agriculture can be achieved.

Yield increase through sustainable agriculture can stimulate agriculture-led economic growth in developing countries. For the world's poor, higher yields lead to lower food prices, food security, and employment generation. Food security will both feed the hungry and provide them with the means to pull themselves out of poverty.

Africa stands to benefit greatly from the new directions planned for agriculture. However, there are still major hurdles to overcome. For example, many European countries ban the importation of food grown from genetically modified seed. As a result, African countries that hope to use biotechnology to grow crops find themselves cut off from their major market supply.

In 2008, genetically modified crops were grown on nearly 300 million acres in 25 countries; 15 of these were developing countries. The world has consumed genetically modified crops since the mid-1990s without incident; GM crops have increased agricultural productivity, increased environmental benefits, and increased farmers' incomes.

It is paramount to move beyond popular biases against the use of agricultural biotechnology; nongovernmental organizations and policy analysts could do much to assist in this process. The nutrition community, using its public credibility and trust, could be an effective voice for reason and act as a feasible counterbalance to the misinformation and fear that still reside in a portion of the public.

The use of transgenic plants offers enormous promise for the rapid integration of improved varieties of traditional crops for developing countries. Clearly, biotechnology is not the "magic bullet" that can free the world from poverty, hunger, and malnutrition; however, it remains a powerful and essential tool, along with plant breeding and innovative agricultural practices such as precision farming, to contribute in a substantial manner toward attaining food security for all.

CHAPTER 7

The Future

Is there a way to produce plant-based food and medicines inexpensively and in greater abundance without damaging the environment or lessening biodiversity? There is an intimate connection between environmental conservation and the utilization of plants for human health. This book has demonstrated the value of identifying natural compounds in plants which in turn serve as leads for the development of new medicines. Biologically active substances are routinely identified from plant tissues. Since plants can also be utilized as inexpensive bioreactors for generating vaccines and pharmaceutical proteins, they offer hope to those who have little to no access to medicine and who can least afford them. Conservation of biodiversity is crucial for the successful identification of novel medicines, and a wide assortment of plant species will continue to play a role in this process. It is a most unfortunate reality that many of the countries that harbor the greatest levels of biodiversity also tend be the least developed (Figure 7.1a). Countries such as these have weak or nonexistent legal frameworks for protecting indigenous knowledge of medicinal plants and fending off biopirates or others who could exploit the natural resources of a region for their own economic benefit. It is no surprise then, that many hot spots of biodiversity throughout the world also overlap with regions of high population density and poverty (Figure 7.1b). The people who reside in these areas are least prepared to sustainably manage natural resources and, in turn, enhance the risk of reducing the level of biodiversity further.

The concept of conserving biodiversity is a key theme of this book, and its importance reverberates throughout its pages. Preserving our sources of natural biodiversity is a critical factor in the context of discovering new medicines. The identification of new traits in the wild relatives of current crops also will help us prepare to feed a burgeoning global population, particularly under the added stress of climate change. Our basic requirement for biodiversity comes full circle, as new varieties of food crops are needed which can produce higher yields on less arable land, so that our wild places, our natural sources of biodiversity and planet's large reservoir of genetic material, can flourish unspoiled.

This final chapter will describe how our ongoing relationship with plants can help us to address our future health concerns. The development of innovative nutrient delivery systems such as nanotechnology to optimize food for a healthier nutritional composition is discussed. The potential of vertical farming, a possible solution for population increases predicted for many cities, is described, as is the present challenge of introducing much needed new crop varieties through agricultural biotechnology to regions such as Africa. Finally, and perhaps most important, the role of public perception in shaping the future impact of plants on human health will be addressed.

FOODS AS DRUGS

When is a plant considered to be more of a food source and when is it more of a drug? Garlic, for example, is mentioned in several chapters of this book in the framework of both medicine and food. It has been suggested that garlic possesses similar properties to aspirin with respect to reducing platelet aggregation; however, unlike taking aspirin tablets, it is extremely difficult to determine precisely how much garlic added to a dish will actually be consumed in a meal, and even more so the amount that was consumed in the context of one's entire diet. Differences in eating patterns, meal preparation, and interactions with other ingredients which could impact the bioavailability of the active compounds in garlic make it next to impossible to be assured of its actual impact in a complex diet.

The very definition of functional food may increase in significance as the world population swells to 10 billion over the next few decades. Along with the increase in population is an expected shift in the proportion of the wealthy; more people of the world will expect to eat better. The increase in the number of wealthy people could mean an increase in what are now considered to be diseases of Western society, including obesity, diabetes, cardiovascular diseases, and cancers. How can we best prepare for these changes?

The ample evidence supporting the beneficial effects of a plant-based diet is one good approach. This book has discussed the importance of phytochemicals which, among other things, play a role in the prevention of chronic diseases. While the health effects of plant antioxidants have been much lauded, in many cases the actual bioavailability of these plant compounds appears to be

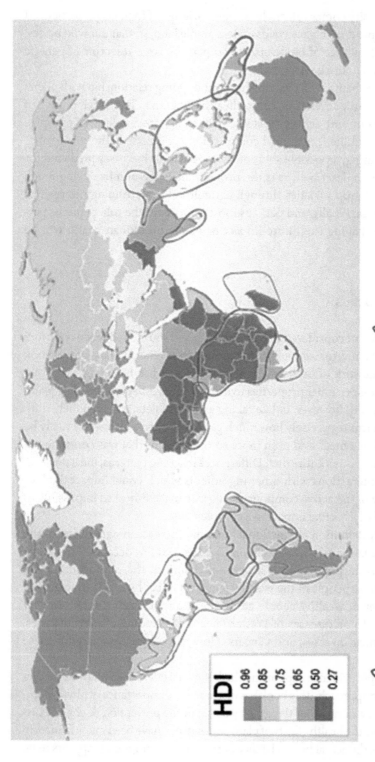

Selected terrestrial biodiversity hotspots   ⬭ Selected major wilderness areas

HDI
| 0.96 |
| 0.85 |
| 0.75 |
| 0.65 |
| 0.50 |
| 0.27 |

Figure 7.1.
(a) Global development and biodiversity. Some of the world's least developed countries are located in hot spot areas of high importance for biodiversity. This map displays the Human Development Index (UNDP) by country and hot spot regions overlaid on that. © 2006 UNEP/GRID-Arendal.

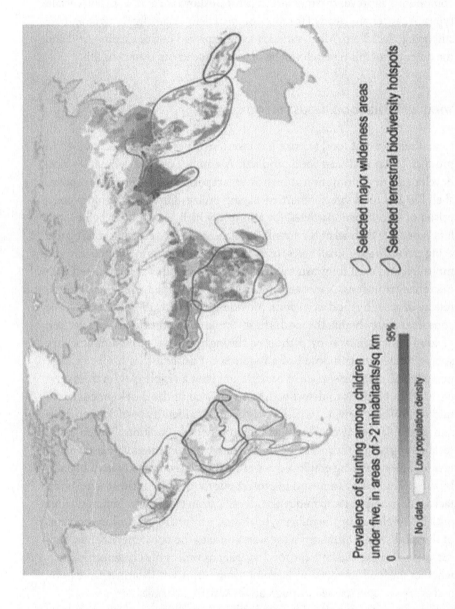

Prevalence of stunting among children
under five, in areas of >2 inhabitants/sq km

0   95%

No data   Low population density

◯ Selected major wilderness areas

◯ Selected terrestrial biodiversity hotspots

Figure 7.1.
(b) Global poverty and biodiversity areas where high poverty and high population density coinciding with high biodiversity may indicate areas in which poor people likely have no other choice than to unsustainably extract resources, in turn threatening biodiversity. © 2006 UNEP/GRID-Arendal.

far too low to play any kind of direct role in preventing disease. A number of researchers have suggested that plant-derived antioxidants may possess other health benefits which in general have a protective effect on the gastrointestinal tract, and this in turn may help to fend off diseases. The precise roles of plant compounds as preventive measures against cardiovascular disease and certain types of cancer are clearly far more complex than originally thought and are still being elucidated. Much remains to be explored in this exciting field and the outcome of this research will be of great value to our overall health.

WHAT WILL THE FOOD INDUSTRY LOOK LIKE?

One direction that food science has taken incorporates nanotechnology to produce better and safer food products. A number of food companies have developed a new innovation known as "smart packaging" to improve product shelf life and food safety. Smart packaging utilizes films or wrappings comprised of silicate nanoparticles, for example, which are stronger, lighter, and less expensive than what is currently available. Materials used in smart packaging can self-repair small holes/tears, respond to changes in temperature and moisture, and even forewarn the customer if the food is contaminated with microbial pathogens. Nanosensors have been developed which react to the release of gases by food as it spoils, providing a signal visible to the would-be consumer as to whether the food is fresh. Some companies focus on detection of food contamination by pathogens through the use of bioluminescence sprays, which are composed of a luminescent protein that has been engineered to bind to the surface of bacteria and emit a visible glow in their presence. DNA biochips to detect pathogens are also under development as are microarray sensors, which will not only be used to identify pesticides on fruits and vegetables but also will monitor environmental conditions in the agricultural setting where they were grown and harvested. Further diagnostic nanotechnologies are being employed not only by the food industry itself but also by the US military, as a means to protect our food supply against terrorist attacks. For example, radio frequency identification (RFID) technology, while initially developed by the military, has found a future role in various portions of the food supply chain, from the warehouse to the consumer's hands. Sensors such as these will help the food supplier as well as the consumer identify whether the food contains any harmful compounds, which farm it came from, and every facility it passed through on the way to the consumer.

Other nanofoods being developed will be interactive, enabling consumers to further customize their eating experience through the use of flavor or nutritional enhancers. These enhancers, present in thousands of nanocapsules, would remain inactive within food and only would be released and activated

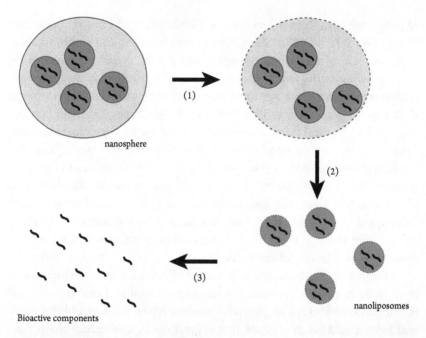

nanosphere

nanoliposomes

Bioactive components

Figure 7.2.
Release of nutrients from encapsulated microspheres (light blue). (1) Microsphere membranes dissolve and release their contents upon exposure to low pH in the stomach, for example. (2) Over an extended period of time, the nanoliposomes (dark blue) release bioactive components (red). (3) Active ingredients could include lycopene, β-carotene, or other phytochemicals.

when triggered by the consumer (Figure 7.2). These nanocapsules can readily enter the bloodstream from the gut, providing superior bioavailability for the nutrient in question. "Smart foods" in the pipeline will be able to liberate their nutrient payload in response to deficiencies that would be detected by nanosensors, also present in the food product. Nanofoods can enhance the nutritional quality of food and improve the way the food is digested.

While the use of nanotechnology to produce safer, nutrient-rich foods is attractive on many levels, it is a reasonable bet that this up-and-coming technology will face the same public perception issues found for genetically modified foods. Lengthy clinical studies will most likely be expected to evaluate the safety of nanofoods that contain additional nutrients before they will be permitted in the marketplace.

AN URBAN ANSWER TO INCREASED FOOD PRODUCTION: VERTICAL FARMS

By 2050, the world population will approach 10 billion, and 70%–80% of the human race will live in cities. Producing enough crops to satisfy the world's hungry will be a significant challenge, as will be accessibility of crops to people who reside in urban centers. Furthermore, the impact of climate change will take its toll in the form of more frequent and severe weather patterns such as hurricanes, droughts, and floods, predictably destroying crops growing in the fields by the millions of tons. Already, scientists are undertaking innovative approaches to tackle this impending problem. One emerging solution can be found in the concept of vertical farming (Figure 7.3). Vertical farms, consisting of multistoried skyscrapers housing fields of crops, are destined to be built in urban centers throughout the world and could provide a sustainable supply of fresh fruits and vegetables without the need to convert more natural environments into farmland. Crops could be generated and harvested year round instead of seasonally and regardless of the weather, thus increasing crop productivity. Transport, storage, and spoilage will be reduced to a minimum, as will the need for fossil fuels used in agriculture today.

A number of drawbacks to the potential of vertical farming exist and will need to be resolved—for example, the requirement and cost for artificial lighting to enable crops to grow and produce fruits and vegetables at optimal levels could be an expense that is great enough to reduce the practicality of such a venture. However, even now designs resembling vertical farms are taking hold in our cities; urban rooftop greenhouses are increasing in number and in popularity in major cities such as New York.

Figure 7.3.
Example of vertical farming. Vincent Callebaut Architectures—www.vincent.callebaut.org.

AGRICULTURAL SUSTAINABILITY AND PROTECTION OF BIODIVERSITY: A DAUNTING CHALLENGE

Many parts of this book have examined the increasing role of biotechnology for generating plants that are advantageous to human health. If any place on Earth has more to gain or lose from the recent advances in agricultural biotechnology, it is Africa. Food insecurity, particularly for sub-Saharan Africa, remains high and will continue to be a major impediment for the progress of the continent for many years to come. Access of the rural poor of Africa to affordable medicines and vaccines through the use of biotechnologies developed in their own countries appears to be a far-off dream. With only approximately 0.5% of its gross domestic product directed toward research and development (with the exception of South Africa, which stands well in the forefront at 0.9% of GDP used for research and development), Africa lags far behind much of the world with respect to generating new crops that will have the capacity to feed its masses. Indeed, continuing negative attitudes toward agricultural biotechnology, specifically of genetically modified crops, have played a role in keeping the continent from reaching its economic potential. In addition to poor financial support, strong policies regarding the regulation and release of GM crops have made it extremely difficult for Africa to advance toward sustainability by any stretch of the imagination. The rate at which GM crops undergo safety assessment and release in Africa is glacial to say the least. As E. Jayne Morris puts it in her recent paper, "The process of gaining regulatory approval for a GMO (genetically modified organism) is not for the faint-hearted."[1] Since the environment in Africa remains unfavorable to the local development of GM crops, there are fears that only large agricultural corporations will have the funding and clout available to take a new transgenic crop through the gauntlet of stringent tests that are required for its regulatory approval. The lack of funding opportunities offered by African governments, coupled with the regulatory red tape that must be overcome to enable these plants to be used by farmers themselves, may backfire for the African people as a whole.

Attempting to feed a vastly growing population using farm practices that produce lower yields, particularly in high-stress environments such as Africa, does not make sense and could very likely result in the necessity to clear even more land for agriculture. Such a strategy could endanger what remains of the continent's rich biodiversity. Chapter 2 described the examination of several African plants as potential medicines to treat HIV/AIDS and malaria. Africa represents a cornucopia of biodiversity, and plants with bioactive compounds that may play a future role in medicine have yet to be identified and examined (Figure 7.4). It is vital that these wild regions, rich in plant species, be conserved.

Figure 7.4.
(a) Africa's biodiversity is remarkably intact. Miombo woodlands in eastern Zimbabwe.
Source: Y. Katerere.

BIODIVERSITY AND THE SEARCH FOR NEW MEDICINES

Most of our new medicines are derived in some manner from compounds that originate from nature, and plants themselves have traditionally played a significant role in the drug discovery process. A great diversity of organisms has been a tremendous source of bioactive compounds for medicine, and the existing gene pool of wild relatives is essential for producing new crop varieties that are better suited to our changing world. As we are still far from identifying a significant proportion of the species of organisms that exist on the planet, it is tragic that many species unknown to us today are under extreme threat of extinction. Conservation of a given species or its fragile ecosystem is important for maintaining biodiversity, as is appropriate compensation for residents of regions from which bioactive compounds originate.

Since much of the human race does not have access to conventional medicine, the need for accessible and inexpensive medicines will continue to be of great importance. Plant-made biopharmaceuticals, including vaccines, antibodies, and other therapeutic agents, offer a viable solution to this problem. Another solution can be found in the further development, through the advancement of plant tissue culture technologies, of cell lines of plant species known to express bioactive compounds.

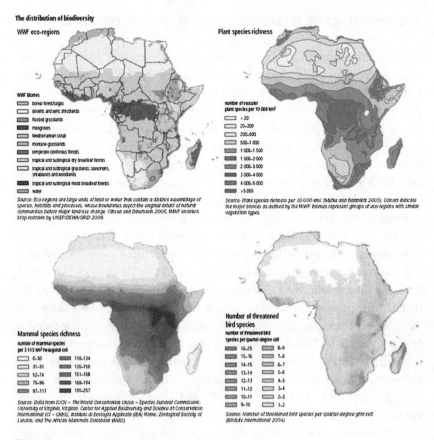

Figure 7.4.
(b) Distribution for biodiversity in Africa.
Source: United Nations Environment Programme; Lead Author, Mark McGinley; Topic Editor, C. Michael Hogan, Ph.D., "Biodiversity in Africa." In *Encyclopedia of Earth*, ed. Cutler J. Cleveland (Washington, DC: Environmental Information Coalition, National Council for Science and the Environment).

Such a technique has been instrumental for the production of Taxol, a compound that not only is derived through harvesting a plant species whose natural habitat has been considered to be endangered but is also difficult and expensive to synthesize in the lab. The original source plant of the bioactive compound can be grown in large cell cultures to produce commercial amounts of the compound. Alternatively, cell lines derived from other more routinely grown plant species which have been engineered to harbor a particular biosynthetic pathway could also produce the compound in demand. Either of these techniques could result in the production of greater quantities of inexpensive medicines and phytochemicals than are available today, without endangering our wilderness areas.

These future strategies for the use of plants to produce food and medicines more effectively through biotechnology sound good in principle but are much more difficult to implement given the current attitudes and beliefs prevailing among the general public. Public perception plays an enormous role in the use of biotechnology to improve our medicine and food supplies. It is no surprise to find a certain degree of commonality between both partakers of herbal medicines and organic food enthusiasts. Both groups may harbor a similar vision of a time when life was less complicated, before factory farms and large pharmaceutical companies entered center stage during the second half of the 20th century. Both groups also appear to have an unrealistic perception of risk assessment. The popular perception that organic food is somehow superior may actually result in a lowered caution in the way some people prepare their food, with the end result being massive food poisoning incidents such as the recent European organic bean sprout outbreak during the summer of 2011. Another classic example is the misperception that organic farming practices must be better for humans and for the environment. Chronic exposure to rotenone for example, a certified organic pesticide, has been shown to cause liver and kidney damage. Similarly, some perceive that herbal medicines are more "natural," and thus safer to use than conventional drugs. In fact, as described earlier in this book, many herbal remedies have proven to be dangerous, either when taken alone or in conjunction with other medicines. A significant percentage of herbal medicines sold today are contaminated with other compounds, as shown in the high levels of heavy metals found in some Ayruvedic medicines mentioned in Chapter 3.

Many Europeans hold the cultivation of genetically modified crops and their entry into the food system under great suspicion (Figure 7.5). A great deal of discomfort lies in the worry that our accessibility to food and medicine may become controlled by a small number of large corporations, who some fear may have more control over our health choices than they should have. The reality is that many large agribusinesses are actively involved in philanthropic activities to help food production in locations such as sub-Saharan Africa, as well as strategic alliances with nongovernmental organizations to develop a sustainable economy in these regions. These same multinational corporations also license their technologies to smaller local companies, providing farmers with a choice of what types of seeds they wish to buy and from whom. In some instances, the technology and expertise is provided from these companies to African scientists on a royalty-free basis.

Figure 7.5.
Stop GMO. Anti-government Solidarity demonstration on June 30, 2011 in Warsaw, Poland. Photo by Tomasz Bidermann / Shutterstock.com.

REGULATING AGRICULTURAL INNOVATIONS CAN BE A DOUBLE-EDGED SWORD

As mentioned in the second chapter, efforts have been made to protect developing countries from acts of biopiracy or commercial exploitation of their natural biodiversity. The formation of the Convention on Biological Diversity was one tactic with the intent to assist in this process. As an unfortunate result of this well-intentioned set of regulations, it is now much more difficult for scientists from universities or research institutes to gain access to the same biological material that was exchanged openly in the past from these same countries. Thus, the attempt to protect the peoples of developing countries from acts of biopiracy of their natural resources has resulted in bottlenecks for researchers in universities and publicly funded institutes, making it difficult for essential scientific research to proceed which could, for example, lead to the next medical breakthrough. In fact, it is the large multinational corporations who can best overcome the bureaucratic hurdles that have now been set in place. As mentioned before in the discussion of overly strict regulations regarding GM crops in Africa, it is becoming exceedingly difficult for any smaller research group or institute, no matter how philanthropic in intent, to achieve its goals.

While clearly some regulations in agricultural biotechnology are necessary, it is a shame that these endeavors to restrict and control innovations in plant research to further improve food crops or identify new medicines appear to be somewhat short-sighted. Efforts to regulate research using biotechnology in such an overly intensive and even antagonistic fashion may inevitably lead to what for many is the worst scenario—that the only groups who are able to move forward in research and development are the very entities that certain factions of the population view with the greatest suspicion: large multinational companies. With this in mind, one can easily imagine numerous confrontations, with publicly funded research scientists, opposition groups, the world's rural poor, and large corporations being the major combatants. Certainly, these four groups do clash from time to time, and the outcomes can be highly unpredictable as seen in the examples provided next.

The use of herbal medicines to treat the enormous HIV-1 positive population of South Africa serves as a first example. Rather than treating their people with antiretroviral medicine, the South African government, led by then president Thabo Mbeki, instead recommended a folk medicine proffered by Mbeki's health minister, consisting of lemons, garlic, and beetroot. This "natural" prescription exacerbated an already dangerous situation and led to an early death toll for Africans suffering from HIV/AIDS that has been estimated to be in the hundreds of thousands. It seemed easier (and more economical to some, perhaps) to embrace an idealistic notion which incorporated traditional beliefs rather than to take a proactive role in combating HIV/AIDS. During this debate, numerous claims were put forward, ranging from charges that scientists were trying to poison African patients to accusations that large pharmaceutical companies were monopolizing the health care and lives of Africans. This ineffectual decision making by the South African government played a catastrophic role in the progression of the immense epidemic of HIV/AIDS in South Africa.

In terms of crops, examples of the clash among large corporations, opposition groups, and the rural poor are rampant. A tumultuous example is found in the story of Bt cotton's introduction to India's farmers. Bt toxin, derived from the bacteria *Bacillus thuringiensis* and originally used as a natural pesticide, has been expressed in GM crops throughout the United States for the past 15 years. Indian farmers were desperate for a new cotton variety that would not succumb to insects or be overly costly in terms of pesticide use. Bt cotton GM plants, grown on test plots in India, were "unofficially removed" by unknown sources and used to interbreed with native crops to produce new Bt varieties. Eventually, hundreds of varieties of "unapproved" Bt-cotton were discovered to have spread across vast sections of Indian farmland in an unregulated fashion. It is interesting to note that in spite of many who condemned these GM crops (some initial attempts to grow test plots of Bt cotton resulted in the crop's destruction by protest groups), plenty of farmers had no qualms about merging a new, useful technology into their own agricultural programs. Today, Bt cotton has revolutionized cotton production in India, resulting in this country's becoming the number one exporter of cotton in the world.

CONCLUSIONS

In spite of the controversies that innovations in plant biotechnology seem to invite, it remains a powerful tool by which food and medicine can be provided to the world's poor. It is important to reemphasize that the majority of the human race is barred from modern medicine as a result of poor accessibility and high cost. As described in this book, plants can be used as bioreactors to produce available and affordable medicines. Biofortified plants could help to address the nutritional needs of the poor and act as a protective measure against infectious diseases. Newly developed functional foods can also provide health benefits and ward off chronic diseases for those in Western industrialized countries.

Plant biotechnology can help to feed the world, reduce arable land use, and maintain biodiversity. A number of reports have already shown that some farmers in developing countries have benefited directly by growing GM crops. Farmers have experienced increases in yields and lower costs of inputs such as pesticides by growing GM crops. This profitability varies from country to country, based on the capacity of the nation to engage in agricultural research and the design of its regulatory infrastructure, among other things. Agriculture is the fundamental driver of economic growth and remains the principal means by which people can raise themselves out of poverty. High-yielding crops will be able to produce more food per acre, thus limiting the amount of wilderness that needs to be converted into arable land. Our wild places (and our reserves of nature's potential new medicines) can be better left intact with the use of biotechnology. As Noel Kingsbury put it in his book *Hybrid*, "There is a powerful environmental argument for sustaining the momentum of the Green Revolution and its high-productivity crops."[2]

More than science is required to make this happen. Improvements must be made in public sector international networks so that these new technologies can be shared and distributed throughout the world on a more equal basis. One hurdle that must be addressed is the handling of intellectual property (IP). While the acquisition of IP does indeed provide a strong basis for motivating research and development in the private sector, it can act as an obstacle for accessibility to new plant varieties and medicines by the poor in the developing world. Another problem faced today is the extreme cost of regulation for the production of new plant-based products.

In the case of food crop production, the end result of an overly strict regulatory framework is that large multinational companies tend to corner the market. Smaller entities from the public sector find it difficult to survive under the regulatory environment that stands today, and portions of the populace remain suspicious about plant seeds and other products originating from large corporations. Changing the regulatory structure to be more science-based and ensuring that the public plays a greater role in the evaluation process of the plant product would not only help publicly funded scientists to realize their goals, but would be a major step forward to helping the poor. Moreover, providing publicly funded scientists better access to plant material from the world's regions of high biodiversity could better ensure the development of new medicines. This must be handled in a manner that adequately addresses the rights and dignity of other cultures, wherever appropriate.

A better informed general public will be a major step in the right direction to providing new foods and medicines that could benefit so many. Although most people know how to operate sophisticated applications on cell phones or computers, they still have a poor working knowledge of where our foods and medicines actually come from. When I spoke at a seminar on some of the topics mentioned in this book at our local public library, it was quite a shock for me to learn how confused and concerned people still are about how biotechnology affects their lives. The general public needs to play a more active role in shaping what our foods and medicines look like in the future. It is very clear that the scientific community needs to reach out and communicate with the public more effectively. This is a challenge in itself, as many scientists find it difficult to free up the time and energy from their own research programs to work on additional projects such as this. It is not a particularly easy feat for many scientists to engage with the general public in layman's terms about their research programs. It took Al Gore to make climate change a household word. Perhaps the world needs a similar champion to provide public awareness of global issues such as food security and biodiversity conservation, before it is too late.

NOTES

CHAPTER 1
1. Reay Tannahill, *Food in History*, p. 367.

CHAPTER 6
1. Robert Paarlberg, *Starved for Science*, p. 20.
2. Michael Pollan, *The Omnivore's Dilemma: A Natural History of Four Meals*, p. 157.
3. Michael Pollan, *The Omnivore's Dilemma: A Natural History of Four Meals*, p. 137.
4. Michael Pollan, *The Omnivore's Dilemma: A Natural History of Four Meals*, p. 247.
5. Pamela Ronald and Raoul Adamchak, *Tomorrow's Table*, p. 25.
6. Michael Pollan, *The Omnivore's Dilemma: A Natural History of Four Meals*, p. 275.
7. Robert Paarlberg, *Food Politics*, p. 145.
8. Robert Paarlberg, *Food Politics* p. 147.
9. Michael Specter, *Denialism*, p. 146.
10. Norman Borlaug, 30th Anniversary Lecture, Norwegian Nobel Institute, Oslo, September 8, 2000.

CHAPTER 7
1. E. Jayne Morris, Modern biotechnology—potential contribution and challenges for sustainable food production in sub-Saharan Africa, *Sustainability* 2011, 3, p. 815.
2. Noah Kingsbury, *Hybrid: The History and Science of Plant Breeding*, p. 416.

REFERENCES

CHAPTER 1

De Vany, Arthur. *The New Evolution Diet: What Our Paleolithic Ancestors Can Teach Us about Weight Loss, Fitness and Aging.* New York, NY: Rodale, 2011.

Harris, Marvin, and Eric B. Ross, eds. *Food and Evolution: Toward a Theory of Human Food Habits.* Philadelphia, PA: Temple University Press, 1987.

Kingsbury, Noel. *Hybrid: The History and Science of Plant Breeding.* Chicago, IL: University of Chicago Press, 2009.

Pollan, Michael. *In Defense of Food: An Eater's Manifesto.* London: Penguin Books, 2008.

Silvertown, Jonathon. *An Orchard Invisible: A Natural History of Seeds.* Chicago, IL: University of Chicago Press, 2009.

Tannahill, Reay. *Food in History.* New York, NY: Three Rivers Press, 1998.

von Nussbaum, F., M. Brands, B. Hinzen, S. Weigand, and D. Häbich. Antibacterial natural products in medicinal chemistry—exodus or revival? *Angewandte Chemie International Editorial England* 45(31) (August 4, 2006): 5072–129.

CHAPTER 2

Boyom, F. F. Antiplasmodial activity of extracts from seven medicinal plants used in malaria treatment in Cameroon. *Journal of Ethnopharmacology* 123(3) (June 25, 2009): 483–88.

Chivion, Eric, and Aaron Bernstein, eds. *Sustaining Life: How Human Health Depends on Biodiversity.* New York, NY: Oxford University Press, 2008.

Gepts, P. Who owns biodiversity, and how should the owners be compensated? *Plant Physiology* 134 (2004): 1295–307.

Goodman, Jordan, and Vivian Walsh. *The Story of Taxol: Nature and Politics in the Pursuit of an Anti-Cancer Drug.* New York, NY: Cambridge University Press, 2001.

Howell, Catherine Herbert. *Flora Mirabilis: How Plants Have Shaped World Knowledge, Health, Wealth, and Beauty.* Washington, DC: National Geographic Society, 2009.

Jeffreys, Diarmuid. *Aspirin: The Remarkable Story of a Wonder Drug.* London: Bloomsbury, 2004.

Mills E., C. Cooper, D. Seely, and I. Kanfer. African herbal medicines in the treatment of HIV: Hypoxis and Sutherlandia. An overview of evidence and pharmacology. *Nutrition Journal* 4 (May 31, 2005):19.

Newman, D. J. Natural products as sources of new drugs over the last 25 years. *Journal of Natural Products* 70(3) (2007): 461–77.

CHAPTER 3

Burgess, J. A., P. F. van der Ven, M. Martin, J. Sherman, and J. Haley, Review of over-the-counter treatments for aphthous ulceration and results from use of a dissolving oral patch containing glycyrrhiza complex herbal extract. *Journal of Contemporary Dental Practice* 9(3) (March 1, 2008): 88–98.

De Smet, P. A. G. M. Clinical risk management of herb-drug interactions. *British Journal of Clinical Pharmacology* 63(3) (2006): 258–67.

Ernst, E. Herbal medicines: Balancing benefits and risks. *Novartis Foundation Symposium* 282 (2007): 154–67.

Houghton, P. J., M.-J. Howes, C. C. Lee, and G. Steventon. Review: Uses and abuses of in vitro tests in ethnopharmacology: Visualizing an elephant. *Journal of Ethnopharmacology* 110(3) (April 4, 2007): 391–400.

Jian-Nuan Wu. *An Illustrated Chinese Materia Medica.* New York, NY: Oxford University Press, 2005.

Li, S., Q. Han, C. Qiao, J. Song, C-L. Cheng, and H. Hongxi Xu. Chemical markers for the quality control of herbal medicines: An overview. *Chinese Medicine* 3 (2008): 7.

Medicina Magica by Hans Biedermann. Birmingham, AL: The Classics of Medicine Library, Division of Gryphon Editions, 1986.

Prakash, S., et al. Lead poisoning from an Ayurvedic herbal medicine in a patient with chronic kidney disease. *Nature Reviews Nephrology* 5(5) (May 2009): 241.

Unschuld, Paul U. *Medicine in China: A History of Pharmaceutics.* University of California Press, 1986.

Yang, Y. F., J. Z. Ge, Y. Wu, Y. Xu, B. Y. Liang, L. Luo, X. W. Wu, D. Q. Liu, X. Zhang, F. X. Song, and Z. Y. Geng, Cohort study on the effect of a combined treatment of traditional Chinese medicine and Western medicine on the relapse and metastasis of 222 patients with stage II and III colorectal cancer after radical operation. *Chinese Journal of Integrative Medicine* 14(4) (December 2008): 251–56.

CHAPTER 4

Clive James Brief 42: Global Status of Commercialized Biotech/GM Crops. Ithaca, NY: International Service for the Acquisition of Agri-biotech Applications (ISAAA), 2010.

Hefferon, Kathleen Laura. *Biopharmaceuticals in Plants: Toward the Next Century of Medicine.* Boca Raton, FL: CRC Press, 2010.

Herbst-Kralovetz, M., H. S. Mason, and Q. Chen, Norwalk virus-like particles as vaccines. *Expert Review Vaccines.* 9(3) (March 2010): 299–307.

McCormick, A. A., S. Reddy, S. J. Reinl, T. I. Cameron, D. K. Czerwinkski, F. Vojdani, K. M. Hanley, S. J. Garger, E. L. White, J. Novak, J. Barrett, R. B. Holtz, D. Tusé, and R. Levy. Plant-produced idiotype vaccines for the treatment of non-Hodgkin's lymphoma: Safety and immunogenicity in a phase I clinical study. *Proceedings of the National Academy of Sciences U S A* 105(29) (July 22, 2008): 10131–36.

Mitchell, Violaine, S. Nalini, M. Philipose, and Jay P. Sanford, eds. *The Children's Vaccine Initiative: Achieving the Vision.* Washington, DC: National Academy of Sciences, Institute of Medicine, Committee on the Children's Vaccine Initiative: Planning Alternative Strategies, 1993.

Sainsbury, F., M. C. Cañizares, and G. P. Lomonossoff. Cowpea mosaic virus: The plant virus-based biotechnology workhorse. *Annual Review of Phytopathology* 48 (2010): 437–55.

Shchelkunov, S. N., and G. A. Shchelkunova. Plant-based vaccines against human hepatitis B virus. *Expert Review Vaccines* 9(8) (August 2010): 947–55.

Verma, D., and H. Daniell, Chloroplast vector systems for biotechnology applications. *Plant Physiology* 145(4) (December 2007): 1129–43.

CHAPTER 5

Christou P., and R. M. Twyman. The potential of genetically enhanced plants to address food insecurity. *Nutrition Research Reviews* 17(1) (June 2004): 23–42.

de la Torre, R. Bioavailability of olive oil phenolic compounds in humans. *Inflammopharmacology* 16(5) (October 2008): 245–47.

Fraser, P. D., E. M. Enfissi, and P. M. Bramley. Genetic engineering of carotenoid formation in tomato fruit and the potential application of systems and synthetic biology approaches. *Archives of Biochemistry and Biophysics* 483(2) (March 15, 2009): 196–204.

Gilani, G. S., and A. Nasim, Impact of foods nutritionally enhanced through biotechnology in alleviating malnutrition in developing countries. *Journal of AOAC International* 90(5) (September–October 2007): 1440–44.

Gomey-Galera, S., et al. Feeding future populations with nutritionally complete crops. *ISB News Report*, January 2010.

Gonzali, S., A. Mazzucato, and P. Perata. Purple as a tomato: Towards high anthocyanin tomatoes. *Trends in Plant Sciences* 14(5) (May 2009): 237–41.

Grant, B. Where's the super food? *The Scientist* 23(9) (September 2009): 30–38.

Iriti, M., and F. Faoro. Grape phytochemicals: A bouquet of old and new nutraceuticals for human health. *Medical Hypotheses* 67(4) (2006): 833–38.

Mayer, J. E. Delivering golden rice to developing countries. *Journal of AOAC International* 90(5) (September–October 2007): 1445–49.

Naqvi, S., C. Zhu, G. Farre, K. Ramessar, L. Bassie, J. Breitenbach, D. Perez-Conesa, G. Ros, G. Sandmann, T. Capell, and P. Christou. Transgenic multivitamin corn through biofortification of endosperm with three vitamins representing three distinct metabolic pathways. *Proceedings of the National Academy of Sciences U S A* 106(19) (May 12, 2009): 7762–67.

Nestel, P., H. E. Bouis, J. V. Meenakshi, and W. Pfeiffer. Biofortification of staple food crops. *Journal of Nutrition* 136(4) (April 2006): 1064–67.

Ortega, R. M. Importance of functional foods in the Mediterranean diet. *Public Health Nutrition* 9(8A) (December 2006): 1136–40.

Pollan, Michael. *In Defense of Food: An Eater's Manifesto*. New York, NY: Penguin Group, 2008.

Ramos, S., L. Moulay, A. B. Granado-Serrano, O. Vilanova, B. Muguerza, L. Goya, and L. Bravo. Hypolipidemic effect in cholesterol-fed rats of a soluble fiber-rich product obtained from cocoa husks. *Journal of Agriculture and Food Chemistry* 56(16) (August 27, 2008): 6985–93.

Ruel, G., and C. Couillard. Evidences of the cardioprotective potential of fruits: The case of cranberries. *Molecular Nutrition and Food Research* 51(6) (June 2007): 692–701.

CHAPTER 6

Ehrlich, Paul R. *The Population Bomb*. San Francisco, CA: Sierra Club/Ballantine Books, 1968.

Clay, J. Freeze the footprint of food. *Nature* 475 (July 2011): 287–89.

Kingsbury, Noel. *Hybrid: The History and Science of Plant Breeding*. Chicago, IL: University of Chicago Press, 2009.

Paarlberg, R. *Food Politics: What Everyone Needs to Know*. New York, NY: Oxford University Press, 2010.

Paarlberg, R. *Starved for Science: How Biotechnology Is Being Kept Out of Africa*. Cambridge, MA: Harvard University Press, 2008.

Pinstrup-Andersen, P., and E. Schioler. *Seeds of Contention:; World Hunger and the Global Controversy over GM Crops*. Baltimore, MD: Johns Hopkins University Press, 2000.

Pollan, Michael. *The Omnivore's Dilemma: A Natural History of Four Meals.* Farmington Hills, MI: Thorndike Press, 2006.

Ronald, P. C., and R. W. Adamchak. *Tomorrow's Table: Organic Farming, Genetics and the Future of Food.* New York, NY: Oxford University Press, 2008.

Specter, Michael. *Denialism: How Irrational Thinking Hinders Scientific Progress, Harms the Planet and Threatens Our Lives.* New York, NY: Penguin Group, 2010.

Thompson, Jennifer A. *Seeds for the Future: The Impact of Genetically Modified Crops on the Environment.* Ithaca, NY: Cornell University Press, 2006.

Thurow, B., and S. Kilman. *Enough: Why the World's Poorest Starve in an Age of Plenty.* New York, NY: Public Affairs Books, 2009.

CHAPTER 7

Choudary, Bhagirath, and Kadambina Gaur. *Cotton in India—A Country Profile.* Ithaca, NY: International Service for the Acquisition of Agri-Biotech Applications (ISAAA), July 2010.

Despommier, Dickson. *The Vertical Farm: Feeding the World in the 21st Century.* New York, NY: Thomas Dunne Books, 2010.

Institute of Food Science and Technology statement on Nanotechnology. www.ifst.org/uploadedfiles/cms/store/ATTACHMENTS/Nanotechnology.pdf.

Kingsbury, Noel. *Hybrid; The History and Science of Plant Breeding.* Chicago, IL: University of Chicago Press, 2009.

Morris, E. Jayne. Modern biotechnology—potential contribution and challenges for sustainable food production in sub-Saharan Africa. *Sustainability* 3 (2011): 809–22.

Morse, Stephen, Richard Bennett, and Yousouf Ismael. Comparing the performance of official and unofficial genetically modified cotton in India. *AgBioForum* 8(1) (2005): Article 1.

Specter, Michael. *Denialism: How Irrational Thinking Hinders Scientific Progress, Harms the Planet and Threatens Our Lives.* New York, NY: Penguin Group, 2010.

ADDITIONAL READING

Chapter 2. Bioprospecting for Medicines from Plants

Clapp, R. A., and C. Crook. Drowning in the magic well: Shaman Pharmaceuticals and the elusive value of traditional knowledge. *Journal of Environment and Development* 11(1) (2002): 79–102.

De Natale, Antonino, Gianni Boris Pezzatti, and Antonino Pollio. Extending the temporal context of ethnobotanical databases: The case study of the Campania region (southern Italy). *Journal of Ethnobiology and Ethnomedicine* 5 (2009): 7.

Dove, A. High-throughput screening goes to school. *Nature Methods* 4(6) (2007): 523–32.

Garrity, G. M., and J. Hunter-Ceera. Bioprospecting in the developing world. *Current Opinion in Microbiology* 2 (1999): 236–40.

Heinrich, M., S. Edwards, D. E. Moerman, and M. Leonti. Ethnopharmacological field studies: A critical assessment of their conceptual basis and methods. *Journal of Ethnopharmacology* 124(1) (2009): 1–17.

Itokawa, H., S. L. Morris-Natschke, T. Akyama, and K. H. Lee. Plant-derived natural product research aimed at new drug discovery. *Journal of Natural Medicine* 62(3) (2008): 263–80.

Kinghorn, A. D., Y. W. Chin, and S. M. Swanson. Discovery of natural product anti-cancer agents from biodiverse organisms. *Current Opinion in Drug Discovery and Development* 12(2) (2009): 189–96.

Levine, D. CEO revives rain-forest drug from failed biotech. *San Francisco Business Times,* July 15, 2005.

Makhubu, Lydia. Bioprospecting in an African context. *Science* 282(5386) (October 2, 1998): 41–42.

MEDUSA Network for the identification, conservation and use of wild plants in the Mediterranean region. http://medusa.maich.gr/.

Miller, K., B. Nellan, and D. M. Sze. Development of Taxol and other endophyte produced anti-cancer agents. Recent Patents in *Anticancer Drug Discovery* 3(1) (2008): 14–19.

Molares, S. Chemosensory perception and medicinal plants for digestive ailments in a Mapuche community in NW Patagonia, Argentina. *Journal of Ethnopharmacology* 123(3) (June 25, 2009): 397–406. Epub March 31, 2009.

Onaga, L. Cashing in on nature's pharmacy: Bioprospecting and protection of biodiversity could go hand in hand. *EMBO Reports* 2(4) (2001): 263–65.

Patwardhan, B., and M. Gautam. Botanical immunodrugs: Scope and opportunities. *Drug Discovery Today* 10(7) (2005): 495–502.

Rajakumar N., and M. B. Shivanna. Ethno-medicinal application of plants in the eastern region of Shimoga District, Karnataka, India. *Journal of Ethnopharmacology* 126(1) (October 29, 2009): 64–73.

Reihling, H. C. Bioprospecting the African Renaissance: The new value of muthi in South Africa. *Journal of Ethnobiology and Ethnomedicine* 27 (2008): 4–9.

Scott, G., and M. L. Hewett. Pioneers in ethnopharmacology: The Dutch East India Company (VOC) at the Cape from 1650 to 1800. *Journal of Ethnopharmacology* 115(3) (February 12, 2008): 339–60.

Sheridan, C. Diversa restructures, raising question over bioprospecting. *Nature Biotechnology* 24 (2006): 229.

Skoczen S., and R. W. Bussmann. ebDB—Filling the gap for an international ethnobotany database. *Lyonia* 11 (2006): 71–81.

Soh, P. N., and F. J. Benoit-Vical. Are West African plants a source of future antimalarial drugs? *Ethnopharmacology* 114(2) (November 1, 2007): 130–40.

Tan, G., C. Gyllenhaal, and D. D. Soejarto. Biodiversity as a source of anticancer drugs. *Current Drug Targets* 7(3) (2006): 265–77.

Traditional healers' practices and the spread of HIV/AIDS in south eastern Nigeria: Herbal medicines for treating HIV infection and Aids. Cochrane Database Systematic Reviews July 20, 2005; (3):CD003937.

Wright C. W. Traditional antimalarials and the development of novel antimalarial drugs. *Journal of Ethnopharmacology* 100 (1–2) (August 22, 2005): 67–71.

Zebich-Knos, Michel. Preserving biodiversity in Costa Rica: The case of the Merck-INBio agreement. *Journal of Environment Development,* 6(2) June 1, 1997: 180–86

Chapter 3. The Lure of Herbal Medicine

Ali, B. H., G. Blunden, M. O. Tanira, and A. Nemmar. Some phytochemical, pharmacological and toxicological properties of ginger (Zingiber officinale Roscoe): A review of recent research. *Food and Chemical Toxicology* 46(2) (February 2008): 409–20.

Bensoussan, A. Establishing evidence for Chinese medicine: A case example of irritable bowel syndrome. *Zhonghua Yi Xue Za Zhi (Taipei)* 64(9) (September 2001): 487–92.

Dasgupta, A. Review of abnormal laboratory test results and toxic effects due to use of herbal medicines. *American Journal of Clinical Pathology* 120(1) (July 2003): 127–37.

Ganzera, M. Recent advancements and applications in the analysis of traditional Chinese medicines. *Planta Medica* 75(7) (2009): 776–83.

Ghayur, M. N., A. H. Gilani, M. B. Afridi, and P. J. Houghton. Cardiovascular effects of ginger aqueous extract and its phenolic constituents are mediated through multiple pathways. *Vascular Pharmacology* 43(4) (October 2005): 234–41.

Gratus, C., S. Damery, S. Wilson, S. Warmington, P. Routledge, R. Grieve, N. Steven, J. Jones, and S. Greenfield. The use of herbal medicines by people with cancer in the UK: A systematic review of the literature. *QJM* 102(12) (December 2009): 831–42.

Grzanna, R., L. Lindmark, and C. G. Frondoza. Ginger—an herbal medicinal product with broad anti-inflammatory actions. *Journal of Medicinal Food* 8(2) (Summer 2005): 125–32.

Isbrucker, R. A., and G. A. Burdock. Risk and safety assessment on the consumption of Licorice root (Glycyrrhiza sp.), its extract and powder as a food ingredient, with emphasis on the pharmacology and toxicology of glycyrrhizin. *Regulatory Toxicology and Pharmacology* 46(3) (December 2006): 167–92.

Karri, S. K., R. B. Saper, and S. N. Kales. Lead encephalopathy due to traditional medicines. *Curr Drug Safety* 3(1) (January 2008): 54–59.

Kiefer, D., and T. Pantuso. Panax ginseng. *American Family Physician* 68(8) (October 15, 2003): 1539–42.

Maheshwari, R. K., A. K. Singh, J. Gaddipati, and R. C. Srimal. Multiple biological activities of curcumin: A short review. *Life Sciences* 78(18) (March 27, 2006): 2081–87.

Moselhy, S. S., and H. K. Ali. Hepatoprotective effect of cinnamon extracts against carbon tetrachloride induced oxidative stress and liver injury in rats. *Biological Research* 42(1) (2009): 93–98.

Ruan, W-J., M-D. Lai, and J-G. Zhou Anticancer effects of Chinese herbal medicine, science or myth? *Journal of Zhejiang University SCIENCE B* 7(12) (2006): 1006–14.

Shakeel, M., A. Trinidade, N. McCluney, and B. Clive, Complementary and alternative medicine in epistaxis: A point worth considering during the patient's history. *European Journal of Emergency Medicine* 17(1) (February 2010): 17–19.

Shukla, Y., and M. Singh. Cancer preventive properties of ginger: A brief review. *Food and Chemical Toxicology* 45(5) (May 2007): 683–90.

Vlietinck, A., L. Pieters, and S. Apers. Legal requirements for the quality of herbal substances and herbal preparations for the manufacturing of herbal medicinal products in the European Union. *Planta Medica* 75 (2009): 683–88.

Xie, P. S., and A. Y. Leung. Understanding the traditional aspect of Chinese medicine in order to achieve meaningful quality control of Chinese material medica. *Journal of Chromatography* 1216(11) (2009): 1933–40.

Chapter 4. Farming Medicines from Plants

Bardor, M., G. Cabrera, P. M. Rudd, R. A. Dwek, J. A. Cremata, and P. Lerouge. Analytical strategies to investigate plant N-glycan profiles in the context of plant-made pharmaceuticals. *Current Opinion in Structural Biology* 16(5) (2006): 576–83.

Bendandi, M., S. Marillonnet, R. Kandzia, F. Thieme, A. Nickstadt, S. Herz, R. Fröde, S. Inogés, A. Lòpez-Dìaz de Cerio, E. Soria, H. Villanueva, G. Vancanneyt, A. McCormick, D. Tusé, J. Lenz, J. E. Butler-Ransohoff, V. Klimyuk, and Y. Gleba. Rapid, high-yield production in plants of individualized idiotype vaccines for non-Hodgkin's lymphoma. *Annals of Oncology* 21(12) (December 2010): 2420–27.

Beyer, A. J., K. Wang, A. N. Umble, J. D. Wolt, and J. E. Cunnick. Low-dose exposure and immunogenicity of transgenic maize expressing the Escherichia coli heat-labile toxin B subunit. *Environmental Health Perspectives* 115(3) (2007): 354–60.

Herbst-Kralovetz M., H. S. Mason, and Q. Chen. Norwalk virus-like particles as vaccines. *Expert Review Vaccines* 9(3) (March 2010): 299–307.

Jin, C., F. Altmann, R. Strasser, L. Mach, M. Schähs, R. Kunert, T. Rademacher, J. Glössl, and H. Steinkellner. A plant-derived human monoclonal antibody induces an anti-carbohydrate immune response in rabbits. *Glycobiology* 18(3) (2008): 235–41.

Kamarajugadda, S., and H. Daniell. Chloroplast-derived anthrax and other vaccine antigens: Their immunogenic and immunoprotective properties. *Expert Review of Vaccines* 5(6) (December 2006): 839–49.

Ko, K., and H. Koprowski. Plant biopharming of monoclonal antibodies. *Virus Research* 111(1) (2005): 93–100.

Lössl, A. G., and M. T. Waheed. Chloroplast-derived vaccines against human diseases: Achievements, challenges and scopes. *Plant Biotechnology Journal* 9(5) (June 2011): 527–39.

McDonald, K. A., L. M. Hong, D. M. Trombly, Q. Xie, and A. P. Jackman. Production of human alpha-1-antitrypsin from transgenic rice cell culture in a membrane bioreactor. *Biotechnology Progress* 21(3) (2005): 728–34.

Naik, G. Teasing vaccines from tobacco. *Wall Street Journal,* February 24, 2010.

Paz De la Rosa, G., A. Monroy-García, M. de Lourdes Mora-García, C. Gehibie Reynaga Peña, J. Hernández-Montes, B. Weiss-Steider, and M. A. Gómez. An HPV 16 L1-based chimeric human papilloma virus-like particle containing a string of epitopes produced in plants is able to elicit humoral and cytotoxic T-cell activity in mice. *Virology Journal* 6 (2009): 2.

Tacket, C. C. Plant based oral vaccines: Results of human trials. *Current Topics in Microbiology and Immunology* 332 (2009): 103–17.

Takagi H., S. Hirose, H. Yasuda, and F. Takaiwa. Biochemical safety evaluation of transgenic rice seeds expressing T cell epitopes of Japanese cedar pollen allergens. *Journal of Agricultural and Food Chemistry* 54(26) (2006): 9901–5.

Thanavala, Y., Z. Huang, and H. Mason. Plant-derived vaccines: A look back at the highlights and a view to the challenges on the road ahead. *Expert Review on Vaccines* 5(2) (2006): 249–60.

Villani, M. E., B. Morgun, P. Brunetti, C. Marusic, R. Lombardi, I. Pisoni, C. Bacci, A. Desiderio, E. Benvenuto, and M. Donini. Plant pharming of a full-sized, tumour-targeting antibody using different expression strategies. *Plant Biotechnology Journal* 7(1) (2009): 59–72.

Chapter 5. Superfoods: Functional and Biofortified Foods

Artajo, L. S., M. P. Romero, J. R. Morelló, and M. J. Motilva. Enrichment of refined olive oil with phenolic compounds: Evaluation of their antioxidant activity and their effect on the bitter index. *Journal of Agricultural and Food Chemistry* 54(16) (August 9, 2006): 6079–88.

Boue, S. M., T. E. Cleveland, C. Carter-Wientjes, B.Y. Shih, D. Bhatnagar, J. M. McLachlan, and M. E. Burow. Phytoalexin-enriched functional foods. *Journal of Agricultural and Food Chemistry* 57(7) (April 8, 2009): 2614–22.

Bouis, H. E. Micronutrient fortification of plants through plant breeding: Can it improve nutrition in man at low cost? *Proceedings of the Nutrition Society* 62(2) (May 2003): 403–11.

Butelli, E., et al. Enrichment of tomato fruit with health-promoting anthocyanins by expression of select transcription factors. *Nature Biotechnology* 26 (2008): 1301–8.

Canene-Adams, K., J. K. Campbell, S. Zaripheh, E. H. Jeffery, and J. W. Erdman, Jr. Relative bioactivity of functional foods and related dietary supplements: The tomato as a functional food. *J. Nutr* 135 (2005): 1226–30.

Cienfuegos-Jovellanos, E., M. Quiñones Mdel, B. Muguerza, L. Moulay, M. Miguel, A. Aleixandre. Antihypertensive effect of a polyphenol-rich cocoa powder industrially processed to preserve the original flavonoids of the cocoa beans. *Journal of Agricultural and Food Chemistry* 57(14) (July 22, 2009): 6156–62.

Cockell, K. A. An overview of methods for assessment of iron bioavailability from foods nutritionally enhanced through biotechnology. *Journal of AOAC International* 90(5) (September–October 2007): 1480–91.

Contaldo, F., F. Pasanisi, and M. Mancini. Beyond the traditional interpretation of Mediterranean diet. *Nutrition, Metabolism, and Cardiovascular Diseases* 13(3) (June 2003): 117–19.

Erlund, I., R. Freese, J. Marniemi, P. Hakala, and G. Alfthan. Bioavailability of quercetin from berries and the diet. *Nutrition and Cancer* 54(1) (2006): 13–17.

Fitó, M., R. de la Torre, M. Farré-Albaladejo, O. Khymenetz, J. Marrugat, and M. I. Covas. Bioavailability and antioxidant effects of olive oil phenolic compounds in humans: A review. *Annali dell'Istituto Superiore di Sanita* 43(4) (2007): 375–81.

Ghosh, D., and A. Scheepens. Vascular action of polyphenols. *Molecular Nutrition and Food Research* 53(3) (March 2009): 322–31.

Haas, J. D., J. L. Beard, L. E. Murray-Kolb, A. M. del Mundo, A. Felix, and G. B. Gregorio. Iron-biofortified rice improves the iron stores of nonanemic Filipino women. *Journal of Nutrition* 135(12) (December 2005): 2823–30.

Hamer, M., and G. D. Mishra. Role of functional foods in primary prevention: Cranberry extracts and cholesterol lowering. *Clinical Lipidology* 4(2)(April 2009): 141–43.

Hotz, C., and B. McClafferty. From harvest to health: Challenges for developing biofortified staple foods and determining their impact on micronutrient status. *Food Nutrition Bulletin* 28(2 Suppl) (June 2007): S271–79.

Howe, P., K. Davison, N. Berry, A. Coates, and J. Buckley. Cocoa flavanols—circulatory and heart health benefits. *Asia Pacific Journal of Clinical Nutrition* 16 (Suppl 3) (2007): S32.

Jeong, J., and M. L. Guerinot. Biofortified and bioavailable: The gold standard for plant-based diets. *Proceedings of the National Academy of Sciences, U S A* 105(6) (February 12, 2008): 1777–78.

King, J. C. Biotechnology: A solution for improving nutrient bioavailability. *International Journal for Vitamin and Nutrition Research* 72(1) (January 2002): 7–12.

King, J. C. Evaluating the impact of plant biofortification on human nutrition. *Journal of Nutrition* 132(3) (March 2002): 511S–13S.

Li, S., F. A. Tayie, M. F. Young, T. Rocheford, and W. S. White. Retention of provitamin A carotenoids in high beta-carotene maize (Zea mays) during traditional African household processing. *Journal of Agricultural and Food Chemistry* 55(26) (December 26, 2007): 10744–50.

Lotito, S. B., and B. Frei. Consumption of flavonoid-rich foods and increased plasma antioxidant capacity in humans: Cause, consequence, or epiphenomenon? *Free Radical Biology and Medicine* 41(12) (December 15, 2006): 1727–46. Epub June 3, 2006.

McKay, D. L., and J. B. Blumberg. Cranberries (Vaccinium macrocarpon) and cardiovascular disease risk factors. *Nutrition Reviews* 65(11) (November 2007): 490–502.

Morris, J., et al. Nutritional impact of elevated calcium transport activity in carrots. *Proceedings of the National Academy of Sciences, U.S.A.* 105 (2008): 1431–35.

Neto, C. C. Cranberry and blueberry: Evidence for protective effects against cancer and vascular diseases. *Molecular Nutrition and Food Research* 51(6) (June 2007): 652–64.

Park, S., M. P. Elless, J. Park, A. Jenkins, W. Lim, E. Chambers IV, and K. D. Hirschi. Sensory analysis of calcium-biofortified lettuce. *Plant Biotechnology Journal* 7(1) (January 2009): 106–17.

Ramiro-Puig, E., F. J. Pérez-Cano, C. Ramírez-Santana, C. Castellote, M. Izquierdo-Pulido, J. Permanyer, A. Franch, and M. Castell. Spleen lymphocyte function modulated by a cocoa-enriched diet. *Clin Exp Immunol* 149(3) (September 2007): 535–42.

Rea, G., A. Antonacci, M. Lambreva, S. Pastorelli, A. Tibuzzi, S. Ferrari, D. Fischer, U. Johanningmeier, W. Oleszek, T. Doroszewska, A. M. Rizzo, P. A. R. Berselli, B. Berra, A. Bertoli, L. Pistelli, B. Uffoni, C. Calas-Blanchard, J. L. S. Marty, C. S. Litescu, M. Diaconu, E. Touloupakis, D. Ghanotakis, and M. T. Giardi. Integrated plant biotechnologies applied to safer and healthier food production: The Nutra-Snack manufacturing chain. *Trends in Food Science and Technology* 22(7) (July 2011): 353–66.

Rudkowska, I., and P. J. Jones. Functional foods for the prevention and treatment of cardiovascular diseases: Cholesterol and beyond. *Expert Review of Cardiovascular Therapy* 5(3) (May 2007): 477–90.

Seeram, N. P. Berry fruits: Compositional elements, biochemical activities, and the impact of their intake on human health, performance, and disease. *Journal of Agriculture and Food Chemistry* 56(3) (February 13, 2008): 627–29.

Stark, A. H., and Z. Madar. Olive oil as a functional food: Epidemiology and nutritional approaches. *Nutrition Reviews* 60(6) (June 2002): 170–76.

Subbiah, M. T. Understanding the nutrigenomic definitions and concepts in the food-genome junction. *OMICS: A Journal of Integrative Biology* 12(4) (2008): 220–35.

Timmer, C. P. Biotechnology and food systems in developing countries *Journal of Nutrition* 133 (2003): 3319–22.

Tucker, G. Nutritional enhancement of plants. *Current Opinion in Biotechnology* 14(2) (April 2003): 221–25.

Welch, R. M., and R. D. Graham. Breeding for micronutrients in staple food crops from a human nutrition perspective. *Journal of Experimental Botany* 55(396) (February 2004): 353–64.

Welch, Ross M., and Robin D. Graham. Genetics of plant mineral nutritional breeding for micronutrients in staple food crops from a human nutrition perspective. *Journal of Experimental Botany* 55(396) (2004): 353–64.

Zhu C., S. Naqvi, S. Gomez-Galera, A. M. Pelacho, T. Capell, and P. Christou. Transgenic strategies for the nutritional enhancement of plants. *Trends in Plant Sciences* 12(12) (December 2007): 548–55.

Chapter 6. Food Security, Climate Change, and the Future of Farming

Baulcombe, D. Reaping benefits of crop research. *Science*: 327(5967) (February 12, 2010): 761.

Beddington, J. Food security: Contributions from science to a new and greener revolution. *Philosophical Transactions of the Royal Society B*. 365 (2010): 61–71.

Bertini, C., and D. Glickman. Farm futures: Bringing agriculture back to U.S. foreign policy. *Foreign Affairs* 88(3) (May–June 2009): 93–10.

Bourne, J. K., Jr. The end of plenty. Special Report: The Global Food Crisis. *National Geographic Magazine* 215(6) (June 2009): 26–59.

Chojnacka, K. Biosorption and bioaccumulation—the prospects for practical applications. *Environment International* 36(3) (2010): 299–307.

Christou, P., and R. M. Twyman. The potential of genetically enhanced plants to address food insecurity. *Nutrition Research Reviews* 17(1) (June 2004): 23–42.

Ejeta, G. African green revolution needn't be a mirage. *Science* 327(5967) (February 12, 2010): 831–32.

Foley, J. Boundaries for a healthy planet. *Scientific American*, April 2010, 54–60.

Gebbers, R., and V. I. Adamchuk. Precision agriculture and food security. *Science* 327 (5967) (February 12, 2010): 828–31.

Godfray, H. C. J., J. R. Beddington, I. R. Crute, L. Haddad, D. Lawrence, J. F. Muir, J. Pretty, S. Robinson, S. M. Thomas, and C. Toulmin. Food security: The challenge of feeding 9 billion people. *Science* 327(5967) (February 12, 2010): 812–18.

Goldstone, J. The new population bomb: The four megatrends that will change the world. *Foreign Affairs* 89 (January/February 2010): 31–43.

Hvistendahl, M. China's push to add by subtracting fertilizer. *Science* 327(5967) (February 12, 2010): 801.

Pennisi, E. Sowing the seeds for the ideal crop. *Science* 327(5967) (February 12, 2010): 802–3.

Pretty, J. Agricultural sustainability: Concepts, principles and evidence. *Philosophical Transactions of the Royal Society B*. 363 (2008): 447–65.

Tester, M., and P. Langridge. Breeding technologies to increase crop production in a changing world. *Science* 327(5967) (February 12, 2010): 818–22.

Timmer, C. P. Biotechnology and food systems in developing countries. *Journal of Nutrition* 133 (November 2003): 3319–22.

Waltz, E. Hungry for GM crops. *Scientific American Worldview (A Global Biotechnology Perspective)*, 2009, 94–95.

Wilcock, R., S. Eliot, N. Hudson, S. Parkyn, and J. Quinn. Climate change mitigation for agriculture: Water quality benefits and costs. *Water Science and Technology* 5(11) (2008): 2093–99.

INDEX

African potato (Hypoxis homerocallidea) 34–37
Agrobacterium 91
Agroinfiltration 96, 97
Algae 157
   Algal Bloom 155
Allergies 101–2
Alkaloids 50
Anthocyanin 116, 122
Anthrax protective antigen 100
Antioxidant activity 36, 58–61, 108–10
   Function 108–10, 111–20, 134, 165, 168
Arsenic 72
Artemisinin, *Artiemisia annua* 15, 33, 174
Aspirin, ASA, salicylic acid 30–32, 165
Ayurveda 56
   Safety of 61–62
Azadirachta 58

Bananas 84, 85
Bayer Innovation GmbH 97, 100
Beta-carotene (β-carotene) 60, 109–10, 120–21, 169
   In Golden Rice 127–28, 131
Bioactive compounds in plants 50–51
BioCassavaPlus 132
Biodiversity, conservation of, biopiracy 5, 15, 16, 37, 141, 161, 177, 178
   Global distribution, "hot-spots of" 164–67
   New medicines 171–73, 178
Biofortified 177
Biofuel 156–57
Biopiracy 15, 164
Biolex Therapeutics 103
Bioreactor 164
Black pepper (Piper nigrum) 60

Book of Plants Al-Dinawari 49
Borlaug, Norman 137, 140–41, 162
Bovine Spongiform Encephalopathy (Mad Cow Disease) 144
Blueberries 119, 120, 122
Bt (Bacillus thuringiensis) toxin 148–50, 176

Calcium 130
Campania 17, 18
Cancer Chemotherapy National Service Center (CCNSC) 28
Cardamom 59
Carrots 130
Caper 115
Carotenoids, in food 109, 110, 114, 120, 121
   In transgenic plants 131
CherryPharm (Cheribundi) 123
Children's Vaccine Initiative 84
Chloroplast 93, 100
Chloroquine 32
Cholera Toxin Subunit B (CT-B) 88, 89
Cholesterol 117
Cinnamon 59
Climate Change 135–37, 164
Clinical Trials 23–26, 51–52
   Of taxol 29–34
   Of herbal medicines 74–84
Cocoa 117–19
CocoanOX 119
Codeine 13, 27
Crofelemer 21
Colchicine 27
Cochlospermum 34
Combinatorial Library 23–24
Compendium of Materia Medica 62
Conservation 164, 172

Convention on Biological Diversity (CBD), and biopiracy 38, 39, 41, 175
Costa Rica 20
Coxlcyclooxygenase-1 (COX-1) 31
Cranberries 119–20, 122
Cryptolepsis 34

Dark Chocolate 118
De Matreria Medica 46
Desertification 136
Dietary supplements 121
Digitalis 26, 41, 51
Dioscordis, Pedanius 46
Diversa 22
DNA Biochip 168
Domestication of Crop Plants 7
Dow AgriSciences 106
Dragon's Blood 22
Drip Irrigation 153
Drug Development 23
Drug Discovery 23–26
Drug Interactions 52–54
Dutch East India Company (VOC) 19

Early Hominids 6
Escherichia coli (E. coli) ETEC 83, 91, 98
    LT-B 91, 98, 102
    in urinary tract infections 119
Edward Stone 30
Elber's Papyrus 30–31, 45
Enquiry into Plants 45
Environmental Protection Agency 105
Ephedra (Ma-huang) 68–69
Ethnobotany 14, 16
    Database 17

Fertilizer 136
    Reduced use 148
    Use in China 153
Food and Drug Administration (FDA)
    As a regulatory body 21–25, 40, 29, 105, 161
    And herbal supplements 69–73
Foxglove (Digitalis lamata) 26, 41, 51
Flavonoids 109, 110, 113, 116
Functional food 107, 108, 112

Garlic 54–55, 114
    Garlic and neem, 58
    Garlic and aspirin 165
Gates Foundation 83

Gene stacking 151
Genetically modified (GM), regulation of 174–75, 177
    Food crops 90
    Carrots 130
    Cotton 176
    Anti-GM 128, 144, 159, 163, 169, 171, 174
    Comparison to organic 161, 163
    And the poor 145
Ginger 59
Ginkgo biloba 55
Ginseng 45, 67–68, 77
Global positioning system (GPS) 151, 154
Glycerrhiza 69–70
Glyphosate 146
Good Laboratory Practice 25
Golden rice 127–28
Green Revolution 137, 140, 141, 163
Greenhouse gas 153
Guiera 34
Gut Associated Lymphoid Tissue (GALT) 88
Gut buster 154

Hairy roots 104
Heavy metals 174
Healing Forest Conservancy 21
Heliobacter pylori 119
Hepatitis B virus (HBV) 84, 85, 91, 97
    HBsAg 97, 103
Heroin 13, 27
Hidden Hunger 123
Hippocrates 30, 45, 55, 79
"Hit" and "Lead" compounds 24
HIV 21, 34–37, 73, 82
    and St John's wort 54
    and garlic 55
    and Africa 171, 176
H1N1 Influenza Virus 97, 99, 105

Icongenetics, Inc. 96
InBio 20, 39
Intellectual Property (IP) 177
Investigational New Drug Application 29
Iron, in crop breeding 124, 126
    Deficiency, biofortification 112, 128–29, 132, 134, 150
Irrigation
    As a modern farming practice 9, 136, 137, 140, 142, 143, 144, 158
    Drip irrigation, microirrigation 153, 154
Irritable Bowel Syndrome 70

Kaglaite 70
Karmataka State 16, 17
Kava 73
 Table 3.1 45
Khoi 17, 19, 34, 35

Leads, in bioprospecting for new medicines
 15, 16, 20, 21, 34, 39, 41, 164
 In drug discovery 23, 24, 25,
 For antimalarial drugs 32, 34, 37
Lead (meavy metal) 61, 72
Levy Muanawasa 144
Licorice (glycerrhiza) 69–70
Lignin 121
Lord Northbourne 158

Madagascar 39
Malaria 31–34, 37, 74, 171
 Antimalarial 15, 27, 32, 34, 38
Malthius, Thomas 137, 140
Mandrake (mandragora) 47–48
Manto Tshabalala-Msimang 37
Marker-assisted selection 150–51
Medicago 99
Mediterranean Diet 110, 112, 113, 114, 116
Merck-InBio agreement 20, 21, 39, 59
Mercury, in herbal medicines 72
Micronutrients 112, 123–26, 130–34, 145,
 150, 162
Minichromosome 149
Milk Thistle 73
Morphine 13, 26
Minerals
 And anti-nutrients 13
 Use in Ayruvedic medicine 56, 61
 In fruit 107, 113
 In biofortified food 125–26, 145, 149
 Deficiencies in soil 129, 132, 133

Nanotechnology 145, 154, 165
 Nanocapsule 154, 168, 169
 Nanosensors 154, 168, 169
Napo Pharmaceuticals 22
National Cancer Institute (NCI) 21, 28, 29
National Center for Complementary and
 Alternative Medicine (NCCAM) 55,
 56, 72, 78, 81
Neem 34, 40, 57–58
Neolithic 6, 7
Newcastle disease virus 106
Nicotine 50

Nimbidin 58
Nimbidol 58
Nitrogen runoff 155
Non Hodgkin's lymphoma (NHL) 100–101
Norwalk virus 83, 84, 88, 98
Nutrigenomics 112

Omega-3-fatty acids 123, 150
Opium 13, 26, 27, 41
Oral Tradition 16, 19, 39, 41
Oregano 115
Organic Farming 158–62, 174

P450 enzymes 36, 52–54
Pacific yew (Taxus brevifolia) 15, 27–30
Paclitaxel 27
Paleodiet 11
Paleolithic 6
Patent medicine 49–50
Pesticide 136, 148–49, 154–55
Peyer's patches 88, 98
Phenols 50
 Polyphenols 116–19
Pharma-Planta Project 86
Pheromone 50
Phytoalexin 111
Phytoremediation 145, 157–58
Picroliv (Picrorhiza kurroa) 75
Pigment 50, 109–11, 122, 127, 148
Plant Breeding 129, 134, 141, 143–145, 150,
 162–63
Plant virus 95–97
Plastid engineering 93–95
Population Bomb, The (by Paul R. Ehrlich)
 140
Precision agriculture 151–52, 163
Proanthocyanidin 119
Prostate Cancer 120–21
Prunus Africana 38

Quinine 27, 31, 32, 34, 41, 43

Rabies 85–86, 91, 98–99, 105
Resveratrol 110, 116–17
Rice, as a staple grain 8, 113, 124, 137, 140,
 149
 Transgenic 91, 93, 101–4, 129, 147,
 148
 Golden rice 127–28
 Biofortified 129–30, 149–50
Risk assessment 174

Rockefeller Foundation 83
Rosemary 115

San 19, 34
Salination, salinity, salt tolerance 91, 136,
    145, 147, 146–48
Secondary metabolite 16, 50, 77
Shamen 14, 37, 39
Shaman Pharmaceuticals 21
Shennong "The Divine Farmer"
    62–63
Shinoga district, Karnataka State
    16, 17
Smart foods 169
Smart packaging 168
Snapdragon 122
St John's wort (Hypericum perforatum)
    36, 51–54
Stress (environmental), stress tolerance 148,
    152, 155
Sub-saharan Africa
    Malaria and 32, 33
    Medical treatment 36
    Cassava 132
    Food insecurity 137, 142, 143, 162, 171,
        174, 175
Sustainable intensification 145–46
Sutherlandia "cancer bush" 34–37
Switchgrass 156–57

Taxol 15, 27–30, 91, 93, 173
Terpenoids 50, 110, 111
Ti plasmid 91
Traditional African medicine 34–37

Traditional Herbal Medicinal Products
    Directive 78
Traditional Chinese Medicine (TCM)
    62–70
Transgenic animals 86
Transgenic plants 89, 90–92
Turmeric 58–59

United Nations 124

Vertical farming 169–70
Vitamins 125
Vitamin A (Beta carotene) 124, 127–28,
    132, 150
Vitamin C (L-ascorbic acid) 108, 110,
    113
Vitamin E 108, 132
Virus-like particles (VLPs) 88, 98
Valerian 27, 55–56

Willow 27, 30–31
Witchcraft Act 40
World Health Organization (WHO)
    and malaria 33
    and herbal medicines 44
    and the Children's Vaccine Initiative 84
    and iron deficiency 128
    and vitamin A deficiency 150
    and pesticide intake 161

Xenotransplantation 86

Zambia 144
Zinc 124, 132, 134, 150